Micro-Enterprises in Agriculture

The Author

Shri M.S. Virdi former Director of CSIR Poly-technology Transfer Centre Bhopal, has served in various National Labs/Institutions of Council of Scientific & Industrial Research (CSIR) and made significant contribution in the field of Research Management, Technology Transfer and Entrepreneurship Development and Rural Development. He also served as Hon. Advisor in Department of Biotechnology of Barkatullah University, Bhopal. He was instrumental in transfer of more than 150 technologies of CSIR in Industrial and Rural Sector. Besides, he made significant contribution in Industrial Development of Madhya Pradesh for which he was felicitated by M.P. Council of Science & Technology, Bhopal. Also, he has received honour and praises from different divisions of CSIR and various other organisations.

In addition to his counselling services to Entrepreneurs he has delivered around 2000 lectures and published over 200 articles and papers to popularize developments in S&T. He has to his credit two books on medicinal and aromatic plants and one book titled "Sustainable Rural Technologies."

Micro-Enterprises in Agriculture

M.S. Virdi

Ex Director, C.S.I.R.
Poly-technology Transfer Centre
Bhopal Ex Natonal Consultant UNDP
Ex Advisor (Hon.)
Barkatullah University, Bhopal

2014

Daya Publishing House®

A Division of

Astral International Pvt. Ltd.

New Delhi – 110 002

Published by   : **Daya Publishing House®**
A Division of
Astral International Pvt. Ltd.
– ISO 9001:2008 Certified Company –
4760-61/23, Ansari Road, Darya Ganj
New Delhi-110 002
Ph. 011-43549197, 23278134
E-mail: info@astralint.com
Website: www.astralint.com

Laser Typesetting   : **Classic Computer Services**
Delhi - 110 035

Printed at   : **Thomson Press India Limited**

PRINTED IN INDIA

Preface

Selection of appropriate product is very important for success of any new venture. Author has offered counselling service to entrepreneurs in product identification and subsequently involved in process of Technology Transfer during last thirty seven years. As a result, nearly 200 units have been established. This book is a step in this direction. The book covers emerging areas like agro-based and food processing, energy, waste utilization, environmental friendly technology, Bio-fertilizers, pollution control, polymers and employment generating technologies which may be useful in converting into enterprises. I am afraid some of these technologies may cross over to SSI sector. The estimates given in the book may marginally increase.

Author is grateful to scientists and Directors of the Institutions and other technology providers for covering their technologies. Author is also grateful to Dr. Neeraj

Sharma for conveying permission of Department of Science and Technology GOI to include author's articles in e-Science Tech Entrepreneur. He is also grateful to Dr. Suresh Malviya for his help and Shri Sudhir Satpute for typing the manuscript. I will be too happy if entrepreneurs make use of informations given in this book.

M.S. Virdi

Contents

POLLUTION CONTROL

POLYMERS

EMPLOYMENT GENERATING TECHNOLOGIES

ORGANIC LEATHER

AGRO BASED ENTERPRISES

1
Manufacture of n-Triacontanol: A Plant Growth Regulator

n-Triacontanol has recently shot into prominence as a plant growth promoter. Successful field trials conducted in U.S.A. have proved its efficacy for higher yield in the case of number of field crops like barley, corn, rice, tomatoes, maize, lettuce, cucumber, etc. Successful field trials of this compound have been conducted by Agricultural Universities who have found it very effective. Yield increases to the extent of 20-30 per cent and a reduction in the dormant shoots (banji) have been reported. Similarly field trials have also been carried out in the tea gardens of North Eastern Region.

n-Triacontanol has been successfully used for other field crops like rice, tomatoes, potatoes, cauliflower, brinjal,

Chilies, ribbed gourd, etc. by some agricultural universities and has been found effective in very low concentrations, *i.e.* 1 to 2 ppm per acre (0.01 to 0.02 mg/acre). The technology has been developed by CFTRI Mysore and transferred to entrepreneurs in Gujarat and U.P. etc. who have set up commercial production units.

Two raw materials used to extract *n*-Triacontanol are:

1. *Tea waste* - Black tea waste including stiff sweepings, tea waste from instant tea processing, damaged tea, decaffeinated tea.

2. *Sugarcane press mud*- Obtained as a waste product during the clarification of sugarcane juice in sugar factories.

Process in Brief

Tea wax/sugarcane wax is extracted from tea waste/ sugarcane press mud by solvent extraction. The wax is fractionated and purified by precipitation. After trans-estrification under appropriate conditions, the compound is further purified and fractionally crystallized with suitable organic solvents. The product is tested for the purity by thin layer chromatography (TLC).

Power requirement	30 H.P.
Manpower employment potential	6 persons

Financial Estimates and Economic Feasibility
(Capacity: 20 gms. of n-Triacontanol/day)

Total project cost	Rs. 0000 (Lakh)
1. Land (1/2 acre)	5.00
2. Building (2000 sq.ft.)	8.00
3. Plant and equipment	15.00

4.	Other fixed assets	2.00
5.	Preliminary and Pre-operative expenses	2.00
6.	Working capital (Margin)	5.00
7.	Total Project Cost	37.00
	Profitability	30 per cent

The concentrated product is diluted and packed in 1 to 2 litre bottles for market use. 0.1 mg (in One litre), is sufficient for one acre of field).

2

Bio-fertilizers

Fertilizers are applied to soil to replenish nutrients like Nitrogen (N), Potash (K) and Phosphorus (P) which is depleted from the soil by each harvest. Chemical fertilizers leave soils degraded, pollute and make it less productive. They also pose severe health hazards. Attempts to increase N- supply by excessive application leads to passage of N - to water bodies, nitrates to ground water and green house gases to the atmosphere. Consumer preferences have shifted towards organic foods grown without the use of chemicals. This awareness is not only growing in developed countries but also in India and other neighboring countries. In India about 56.29 million hectares of land is either wasteland or fallow land which is proposed to be brought under cultivation. Thus total area under cultivation will be 170.29 million hectares. There is assured demand for bio-fertilizers to make agriculture profitable. Bio-fertilizers include nitrogen fixers (symbiotic or non symboitic bacteria), phosphate solubilising fungi and bacteria and mycorrhizal fungi rhizobacteria that transfer the nutrients from soil to the plant roots. Biofertilizers like *Rhizobium, Azotobacter, Azospirilium* and Phosphate solubilising bacteria fungi are used. Carrier materials such as peat, lignite, peat soil, wood charcoal favour the growth of micro-organisms. Microbial inoculants like Azolla, Trichoderma, Frankia and Vesicular Arbuscular Mycorrhiza (VAM) are available in the market.

Micro-organism	Nutrient Fixed
Algae	25kg/N/ha
Azolla	900 N/ha
Azospirilium	10-20 kg N/ha
Rhizobium	50-300 kg N/ha
Azotobacter	Upto 20 kg N/ha
Mycorrhiza	Solubilise good phosphorous (60 per cent)

Government of India (DBT, New Delhi) provides non-recurring grant in aid of upto Rs. 20 lakh for setting up of Biofertilizer production units of 150 MT annual capacity by entrepreneurs, NGOs, industry etc.

In India, present demand of phosphatic fertilisers is 47.98 lakh tonnes and chemical industry is able to supply 33.48 lakh tonne and balance of 14.50 lakh tonnes is imported. Mycrorrhiza provides phosphorus nutrition to the plants. These fungi utilises phosphorus from extremely low concentration and could offset the high cost of phosphate fertilisers. It can at least partially replace chemical fertilizers. For technology one can contact, The

Energy Research Institute (TERI), India Habitat Centre, Lodhi Road, New Delhi 110003. For other Bio-fertilisers one can contact IARI, New Delhi and other Agricultural Universities.

3
Vermicompost

After independence, India faced a problem of food scarcity due to its high population growth. The solution of this problem was one of the main issues before the planning commission. It was felt that this problem can be solved by increasing the agricultural production in the country. Agricultural production could be increased by introducing high yielding variety (HYV) seeds, fertilizers, pesticides, crop rotation, development of irrigation facilities and mechanization. Use of various inputs improved the agricultural production and productivity but simultaneously created environmental problems.

The use of fertilizers and pesticides has proved more harmful to soil as well as crop itself. Intake of these chemicals gave rise to several types of side effect to human beings.

These problems led the scientists and researchers to revise techniques, which would not be harmful to people as well as environment. To overcome the problems to some extent, they have developed various locations and climate to specific technologies.

The development of organic farming is one of the important technologies in this direction. The vermi composting is one of the techniques of natural sustainable farming. These techniques use waste household organic materials of farmyard and agricultural fields as raw materials and produce environment friendly manure for agricultural production. It generates alternative means of earning and it reduces cost of farm inputs.

The chemical analysis of vermi compost shows that it has high level of nitrogen, phosphorus and potassium, which are necessary for growth and development of plants. In the present scenario quantity of solid waste organic material is increasing day by day in rural and urban areas. Earthworms convert all types of waste organic materials into useful manure. They are unable to digest rubber, plastic, metals, stone and glass etc. Due to this reason these materials should be removed from the waste materials before using for this purpose.

Improvement over Traditional Technology/ Process

The farmers are using dung as manure directly without their decomposition in proper manner in the region, which is not giving appropriate result and increases the population of weeds. In this method large proportion of cow dung also get washed away with surface water without improving

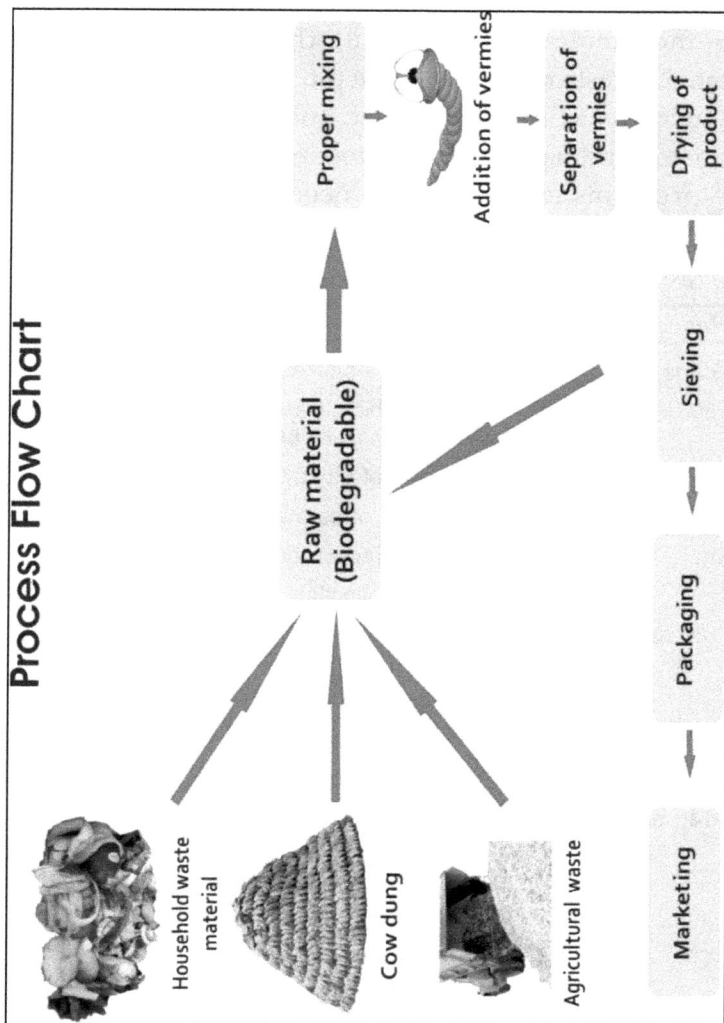

Figure 3.1

the nutrition power of the soil because it is not getting appropriate time to decompose. Sloppy lands constitute major problem of run off.

Some farmers used to put their cow dung and household waste materials in the pit. But due to the lack of moisture for a long period it does not get properly dissociated and not fully converted into useful manure. All these problems emphasize the need to introduce vermi composting techniques for high agricultural production in the region.

Period	Pit	Process
0-30 days	Pit- 1	Collection of biomass and cattle dung
30-60 days	Pit- 1	Soaking of biomass with water, cow dung slurry and covering with black polythene sheet.
	Pit- 2	Collection of biomass
60-90 days	Pit- 1	Inoculation of earthworms
	Pit- 2	Biodung preparation
	Pit- 3	Biomass collection
90-120 days	Pit- 1	Vermicompost ready; migration of earthworms to pit 2
	Pit- 2	Vermicomposting
	Pit- 3	Biodung preparation
	Pit-4	Collection of biomass
120-140 days	Pit- 1	Harvesting of compost and collection of biomass
	Pit- 2	Vermicompost preparation and migration of earthworms to pit 3
	Pit- 3	Vermicomposting
	Pit- 4	Biodung preparation

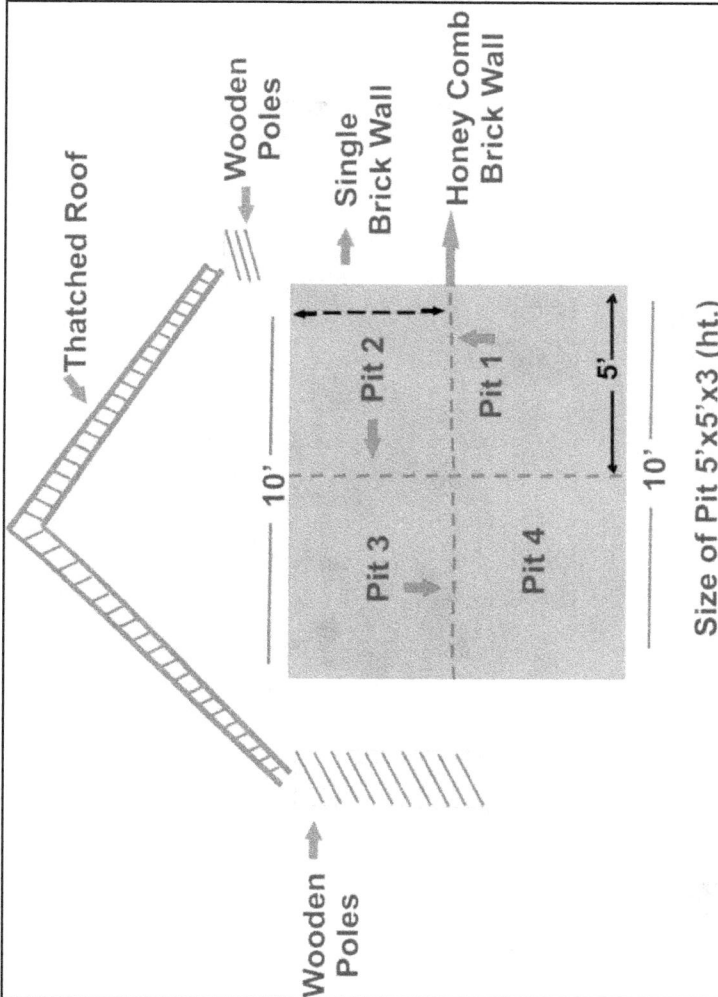

Thatched Roof

Wooden Poles

Single Brick Wall

Honey Comb Brick Wall

Pit 2

Pit 1

Pit 3

Pit 4

10'

10'

5'

Wooden Poles

Size of Pit 5'x5'x3 (ht.)

Figure 3.2

Figure 3.3: Traditional Vermicompost Pit

Techno Economics

(a) Infrastructure	
Cost of shed (35x25ftx875) Sq. ft.=	Rs. 5,000
Cost of water tank 2000 Lit. capacity=	Rs. 3,000
Equipments:	Rs. 5,700
Tasla, Phada etc.	Rs. 500
Sewing Machine	Rs. 3,000
Sealing Machine	Rs. 2,000
Cost of two steel buckets	Rs. 200
Preperation of Beds (10 beds)	Rs. 1,500
Cost of Worms (1500xRs. 0.5x10)	Rs. 7,500
Total	**Rs. 22,700**

(b) Cost of Production (one cycle)	
Cost of cow dung (100kgx10xRs. 0.5) 1000 kg	Rs. 500
Cost of Waste material (200x10xRs. 0.25)	Rs. 500
Bed filling and removing compost labour charge	Rs. 500
Maintenance labour charge (45 daysx1/2hr) 22.5hr/8	Rs.150
Packaging labour charge (4xRs. 50)	Rs. 200
Cost of packaging material	Rs. 2500
Total	Rs. 4350
Total of Nine cycles (Rs. 4350x9)	Rs. 39150

(c) Output		
(i)	Total production of vermi compost	
(ii)	(18kgx10)= 180kgx9 cycles = 16,200 kg @Rs. 3/-	Rs. 48,600
(iii)	Selling of earth worms (60,000xRs. 0.50)	Rs. 15,000
	Total	**63,600**

(d) Profit

Profit will vary in subsequent years

1st yeat rs. 63,600 – (Rs. 22,700+Rs. 39,150)	Rs. 1,750
2nd year Rs. 63,600 – Rs. 39,150	Rs. 24,450
3rd year Rs. 63,600 – Rs. 39,150	Rs. 24, 450
Total	**Rs. 50,650**

The above profit is calculated on the basis of 10 beds. If number of beds increases the percentage of profit will improve due to scale of economy. There is increasing demand of vermicompost in cities for kitchen gardens.

Source of Technology

Centre of Science for Villages (CSV) Sewa Gram, Wardha, Maharashtra, M.P. Vigyan Sabha, Bhopal and many others.

4
Tissue Culture Technology

Tissue Culture technology covers a wide range of techniques including *in vitro* culture of organs (shoot tips, root tips, flowers, ovaries, ovules, embryos, anthers, etc.) tissue cells and protoplasts. Tissue culture has emerged as a valuable tool for mass propagation of several economically important species of medical and aromatic plants, flowers, fruits etc. for generating disease-free stocks, stress tolerant varieties of higher yielding crops and for long term storage of germ plasm. The plant cell tissue culture has generated immense impetus to biotechnological research.

History

The beginning of tissue culture dates back to 1920's when German Botanist Haberlandt initiated culture of leaf cells in Knop's solution. Subsequently, the modern scientific investigations advanced and improved the traditional skills required to propagate *in vitro* and continuing with artificial seed production.

Tissue culture has proved to be a good technique for mass propagation of a number of plants covering horticulture crops, floriculture, forestry species medicinal and aromatic plants which are rather difficult to propagate by conventional methods. Tissue culture was considered to be complicated technology but with passage of time several R&D Institutions and voluntary organizations have made efforts to simplify the procedures and have avoided the use of costly and sophisticated equipment so that educated farmers could adopt the technology at field level by taking 3-4 month training in organizations like NCL Pune, CIMAP Lucknow, NBRI Lucknow, TERI Gurgaon, Institute of Himalayan Bioresource Technology Palampur, COSTFORD, Kerala, ARTI Pune, IIHR Bangalore, etc.

There is no universal technology for the propagation of all plant species. Thus, separate protocols have to be developed for every species. The developed protocols are tested for reproductibility and transferred to persons for propagation. Multiplication is a purely routine activity which can be carried out by any person having some basic educational background. The basic difference in propagation of different species is due to difference in the chemical composition of culture medium used at various stages of growth. A person having acquired basic training in mass propagation techniques can propagate any species after familiarization with that particular protocol.

Following steps are involved in micro-propagation through tissue culture techniques:

Preparation of the appropriate sterile culture medium

↓

Culturing of the selected plant parts after surface sterilization

↓

Manipulation of chemical composition of the medium for modulation of stage of growth

↓

Development of plant/shoots

↓

Rooting of the plant

↓

Hardening to withstand external environment

Supply of plants has now become commercial activity for forestry, horticulture, floriculture and even agriculture, bio-energy plantation etc. Some entrepreneurs are running lucrative business and low cost tissue culture lab is a part of enterprise.

Earlier, Tissue Culture laboratory involved high science and investment which is essential for development of protocols. Two decades back when National Chemical Laboratory developed planning material for bamboo and sugarcane propagation, it revolutionized the concept.

Subsequently, most of bio-science laboratories set up Tissue Culture labs to develop varieties of plant material falling in their scope of R and D. Some NGO's like COSTFORD, Kerala, ARTI Pune and IHBT (CSIR) Palampur set up low cost Tissue Culture labs by replacing costly equipment with simple devices such as Laminar air flow unit which costs around Rs. 1.00 lakh has been replaced by small hood with mica lining and fitted with a germicidal UV lamp. Now, Tissue Culture activities are being carried out in villages. Vasant Dada Sugar Institute, Pune has propagated Sugarcane in lacs of acres in Maharashtra using Tissue Culture.

Clean, pollution free environment of rural areas is congenial for Tissue Culture labs and help in reducing investment. The nutrients and chemicals used should be of LR grade.

Cost of Equipment

Cost of Equipment for a moderate Tissue Culture laboratory is around Rs. 5-6 lakhs.

Cost of Small Laboratory

A small laboratory could be established at a low cost of Rs. 25,000-30,000.

Equipments Required

☆ Laminar air flow

☆ Air shower curtain

☆ Autoclave

☆ Inoculation cabinet

☆ Rotary shakers

☆ Low speed centrifuge

☆ pH meters

☆ Balance

☆ Microscope

☆ Oven

☆ Glass ware

☆ UV lamps

All the equipments are easily available from companies dealing in scientific instruments. Micro nutrients, organic and inorganic chemicals required in growth of tissues and seedlings are also easily available from dealers of chemicals.

All these costly equipment have been replaced by simple tools and devices in establishing rural tissue culture labs.

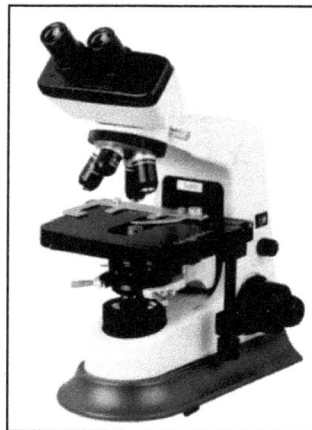

☆ Hood with UV lamp replaces Laminar air flow unit.

☆ High humidity chamber can be made locally by fixing humidifier.

☆ Pressure cooker and stove replaces autoclave used in conventional process, temperature and humidity control are achieved in green house for hardening the plants which can be done in make shift poly houses set up in the shade of trees.

Other items include balance, measuring vessels, glassware and miscellaneous items like knife, forceps, spirit lamp, cotton, hand sprayers etc. As regards chemicals once protocol is developed repeatability is achieved.

Almost all Agricultural Universities have set tissue culture labs. Entrepreneurs can approach them.

Source of Technology and Training

☆ Institute of Himalayan Bioresource Technology Palampur-176061 H.P.

☆ The Energy Research Institute (TERI), Gurgaon, Haryana.

☆ COSTFORD (Biotechnology Unit) Cherthekulangara, Mavelikara - 690106, Kerala.

☆ Appropriate Rural Technology Institute (ARTI) 11, Maninee Apartments, Survey 13, Dhayari Gaon, Pune 411041, MH.

☆ CIMAP, Lucknow.

Success Story

Rural Women, Modern Techniques

☆ Department of Biotechnology, Government of India sponsored a project to be implemented by the Institute for undertaking the challenging task of capacity building of rural women in the modern techniques of plant tissue culture which was hitherto considered a very sophisticated technology. The Institute selected two groups comprising 25 women each hailing from two different locations in the Kangra district of Himachal Pradesh.

☆ In the first phase these groups were familiarized with the activities of IHBT in the field of Tissue Culture and they were educated about the advantage of plants raised through Tissue Culture. The whole technology was published in the form of manual in Hindi covering text and illustrations. Subsequently, a mobile laboratory fitted with a custom made Laminar air flow with a backup generator set was pressed into service and scientists visited these two villages periodically to train women in the process of Tissue Culture starting from media preparation, autoclaving, explanation to inoculation under aseptic conditions. Realising the need of reducing contamination because of movement of culture to their homes which were 25-25 km away from place of work two separate units complete with Laminars, culture racks and mini-auto-claves were setup. This enhanced the success rate of sterile cultures. The women groups hired rooms in the villages and shared the rent, electricity bill and water charges. A small hardening facility in the form of poly house was established for stabilisation of plants raised through Tissue culture.

☆ These women groups are now proficient enough to work independently and are concentrating on high floriculture. Such efforts can be replicated elsewhere in the country and tissue culture will no longer remain within the four walls of high tech laboratory or industry.

☆ An initiative of IHBT, Palampur, Himachal Pradesh

5
Papain from Papaya

Papaya is a versatile plant having number of uses and enzymatic properties. It could also be called medicinal plant rendering high income to farmers. One can manufacture products like tutti-frutti, candy slices (*murabba*) and dehydrated papaya powder from raw papaya.

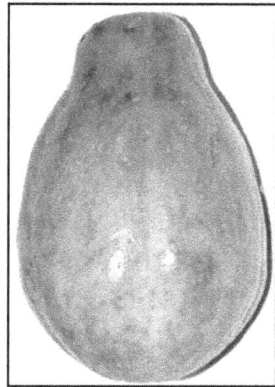

Origin of Papaya

Though the exact area of origin is unknown, papaya is believed to be a native to tropical America, perhaps in southern Mexico and neighbouring Central America. It is recorded that seeds were taken to Panama and then the Dominican Republic before 1525 and cultivation spread to warm elevations throughout South and Central America, southern Mexico, the West Indies and Bahamas and to Bermuda in 1616. Spaniards carried seeds to the Philippines in 1550 and the papaya travalled from there to Malacca and India. Seeds were sent from India to Napal in 1626. Now the papaya is familiar in nearly all tropical regions of the World and the Pacific Islands and has become naturalized in many areas. Up to about 1959, the papaya was commonly grown in southern and central Florida in home gardens and on a small commercial scale.

The Latex of the papaya plant and its green fruits contains two proteolytic enzymes, papain and chymopapain. The latter is most abundant but papain is twice as potent. In 1933, Ceylon (Sri Lanka) was the leading commercial source of papain but it has been surpassed by East Africa where large-scale production began in 1937.

Papain

Papain is the only plant proteinase now being used as a protein digestant. It is used in combating dyspepsia and other digestive disorders. It is used in the making of pharmaceutical preperations like glyc. papain, elix papain, liq. papain and liver tonics. Papain finds extensive use in the making of proteolysed preperations of meat, liver and casein. In brewing industries it is also used in tenderising meat and softening of leathers, degumming of natural silk and wool fabrics, chewing gums, tooth paste and cosmetics. Apart from indigenous demand its 80-90 per cent production is exported to USA and European countries, Western countries import purified papaya latex for further processing to meet their specialized needs.

Cultivation of papaya is suitable in tropical and sub-tropical regions. In India it is cultivated in large areas. Fertile, water retaining soil is most suitable for its cultivation. Tamil Nadu Agriculture University has developed CO and CO_2 varieties of papaya especially for Latex of papain production. Farmers can cultivate papaya for production of latex in about 25 acres (10 hectares) of land for supply 50-60 kg of latex per day for 300 days. In order to get papain throughout the year the plantation is done in three phases with a gap of 3-4 months.

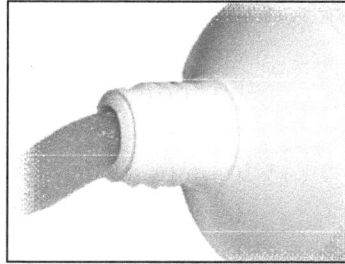

Uses of Papain

Papain by far is the most widely studied of the cysteine enzymes because of its commercial value. It is used in applications like tenderising of meat for fast cooking, softening of silk fibre and leather, chill proofing of beer and pharmaceutical preparations for curing stomachic ailments. Papaya pulp is also used in food processing industry. Some other uses of papain include:

☆ Defibrinating wounds in hospitals

☆ Used in pet food to reduce viscosity and increase palatability

☆ Shrink proofing of wool

☆ Prevents cornea scar deformation

☆ Used to treat edemas, inflammatory processes, and in the acceleration of wound healing

☆ It is used as an ingredient in cleaning solutions for soft contact lenses

☆ In low doses it can be used as an indigestion medicine.

☆ Papain, in the form of a meat tenderizer such as Adolph's, made into a paste with water, is also a home remedy treatment for jellyfish, bee, yellow jacket (wasps) stings, mosquito bites, and possibly stingray wounds, breaking down the protein toxins in the venom.

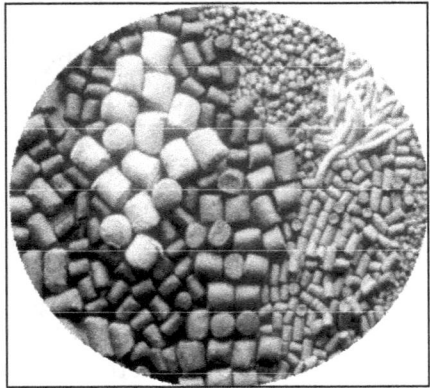

☆ It can also be found as an ingredient in some toothpastes or mints as teeth-whitener

☆ Papain has been employed to treat ulcers, dissolve membranes in diphtheria, and reduce swelling, fever and adhesions after surgery.

Production of Papain

Farmers generally cultivate CO, CO_2 varieties developed by TN Agriculture University and Taiwanese varieties. One plant gives a yield of about 100-150 kg of fruit in 8-10 months period. Plant to plant distance of 2 meters is kept. Some farmers adopt inter-cropping practice and grow chillies for higher return. One can generate a profit of 10-12 lakhs from a plantation of 20 acres. A small plant for production of papain could be set up near the field which costs around Rs. 6.00 lakhs on machinery apart from infrastructure like shed, utilities like water and electricity.

Process of Papain Production

The unripe papaya is first given a longitudinal cut in the evening and it is wrapped by polythene bag for collection of latex during next day morning with the rising of sun. After collection the latex is mixed with potassium meta by-sulphite (Kms) and kept in cold storage to avoid fermentation. Subsequently the latex is spread in trays and drying takes about 4 hours. After drying the latex is scratched from the trays. It is again mixed with Kms and ground in SS hammer mills or roller mills to get fine powder. Then its proteolactec activity is tested in laboratory. Subsequently lactose is added to the latex powder to get BP or IP grade papain. The ready papain is finally packed in airtight polythene bags or metal containers.

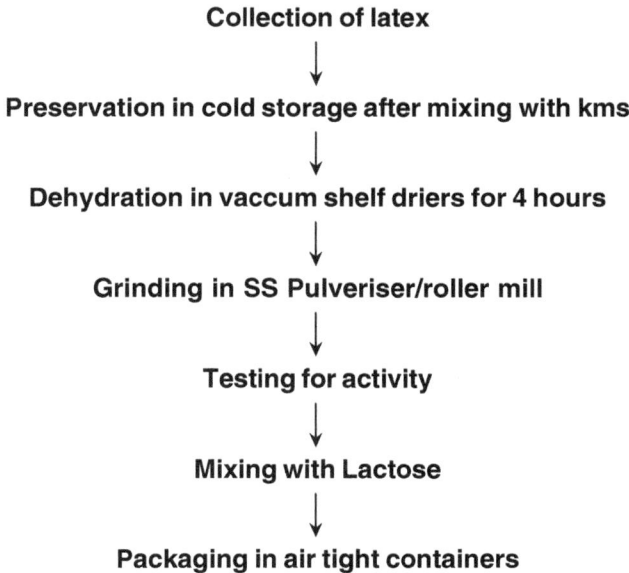

Collection of latex

↓

Preservation in cold storage after mixing with kms

↓

Dehydration in vaccum shelf driers for 4 hours

↓

Grinding in SS Pulveriser/roller mill

↓

Testing for activity

↓

Mixing with Lactose

↓

Packaging in air tight containers

Process Flow Chart

The main requirements of equipment are vacuum shelf drier, de-humidifier, hammer mill, blender, lab equipment and auxiliary equipment like cans, weighing scale, walk-in cooler sealing machine.

All this equipment is available indigenously from food processing machinery manufacturers.

Techno-Economics

Capacity- 5000 kg of latex papain 1 year	
Land – 400 sq metres	0.50 lakhs
Building 100 sq metres	2.00 lakhs
Machinery	6.00 lakhs
Contingency and pre-operators	1.00 lakhs
Misc.	Rs. 9.50 lakhs+ 1.00
Working capital (without raw material)	**Rs. 10.50**

Source of Technology

Technology for papain production could be obtained from CFTRI, Mysore.

The plant should be set up near the field but water and electricity should be available. In addition, farmer can sell the ripe fruit in the market. It is a highly profitable venture.

6

Processing of
Medicinal Plants

The use of medicinal plants for health care is as old as human civilization. The oldest repository of human knowledge is 'R*igveda* written in era 4500-1600 BC. Later '*Atharvaveda'* mentions about 800 formulations used in '*Ayurveda'*. CSIR has digitalized this knowledge base for wider applications throughout the world known as TKDL. In the 20th century there was perceptible change from herbal medicines to synthetics due to R and D in western world but herbal medicines continued to cater to the needs of Third World countries. There has been renewed interest in herbal drugs due to their unique properties and cheaper cost and thus bringing them within the reach of common man and Organizations like WHO have recognized the potential of folklore/ethno system of health care and included herbal drugs in "National Health Care Programmes". In global context it is necessary to define the term medicinal plant, In broader sense all herbal plants have been credited with having medicinal properties - effects that relate to health or which have proven to be useful as drugs by western standards or which contain constituents that are used as drugs. It is estimated that there is Rs. 40,000 crores world market for medicinal and aromatic plants which is likely to grow due to low toxic effects of medicinal plants. There is export of above Rs. 4000 crores of medicinal plant from India.

A large quantity of medicinal and aromatic plant raw-materials have been collected from the forest sources but due to rise in demand cultivation has been taken up in various parts of the country. Earlier traditional suppliers and traders dealt with the industry. Now, CIMAP and others have published information for the benefit of growers who can now directly supply plant material to pharmaceutical industry. National Medicinal Plant Board (NMPB), New Delhi, National Horticulture Board, NABARD and other nationalized banks provide soft financial assistance for cultivation of medicinal plants.

In alternate system of medicine very large variety of plant materials is used for ayurvedic formulations. Medicinal plants, those are required for commercial supply to pharma companies are shown in Table 6.1.

Processing involves Two Methods

1. Drying process.
2. Extraction Process.

Table 6.1

Sl.No.	Name of Plant	Drug	Use
1.	*Acorus calamus* (Vacha)		Tranquiliser
2.	*Adhotoda vasica* (Vasa)		Oxytocic
3.	*Andrographis paniculata* (Kalmegh)		Bacillary dysentry
4.	*Asparagus racemosus*-satavar		Anti-oyxtocic, anti-ulcer, lurv and treatments of burns wound healing enhance lactation.
5.	Belladonna		In peptic ulcers, intestinal disorders, colic and urinary bladder pains, kidney stones.
6	Cinchona		quinine-antimalarial property
7	Datura		Source of tropane alkaloids
8	Digitalis		Digitoxin-congestive heart failure, digitalis-cardiac muscles.
9.	*Discoria composita floribunda*		Diosgenin precursors for synthesis of cortisones, sex hormones and oral contraceptives.
10.	Ergot (Ergot of rye)- (fungus on rye plant)		Alkaloids, analgesic for migraine (ii) Hypertension and other peripheral circular disorders.
11.	Hyoscyamus (Henbane)		Relieve spasm of urinary tract with strong purgative to prevent gupping.

Contd...

Table 6.1–*Contd...*

Sl.No.	Name of Plant	Drug	Use
12.	Liquorice		Flavouring agent, demulcent and mild expectorant, anti-spasmlytic and anti ulcergenic.
13.	Opium poppy		Morphine-relieves from extreme pain-codeine-analgesic papaverine-smooth muscle relaxant in asthamatic, gastric and intestinal spasms and coronary ailments.
14.	Periwinkle (*Catharanthus roseus*)		Anti-cancer drugs-vincristine-vinablastine.
15.	Psyllium seed		Chronic constipation, also controls diarrhoea and dysentery. Seed mucilage used as stabilizer in ice creams, chocolates and other food materials.
16.	Artemisia annua		Two derivatives Artemisinin and arteether-very effective against chloroquine resistant strains of malaria parasite as well as against cerebral malaria.
17.	*Rauvolfia serpentina*		Sedative-alkaloides-reserprine and vescinnanins are used in modern medicine as transquiliser and hypertensive.
18.	*Senna-Casia Augustifolia*		Laxative
19.	(i) *Bacopa monnieri* (Brahmi)		Anti-anxiety agent improves intellect and memory.
	(ii) *Centella asiatica*		Vulnerary wound healing.
20.	*Casia papaya* (Papita)		Proteolytic mucolytic.

Contd...

Table 6.1–Contd...

Sl.No.	Name of Plant	Drug	Use
21.	*Curcuma longa* (Haldi)		Choleretic and other usages
22.	*Picrorhiza kurroa* (Kutki)		Anti-hepatotoxic.
23.	*Coscinium Fenestralim* (Dam Haldi)		Cholera and gastroenteritis
24.	*Taxus baccata*		Anti-cancer drug (Taxol)
25.	*Glorisoa superba*		Colchicine
26.	*Mappia foetida* (Ghaneru)		Anti cancer
27.	*Withania somnifera* (Ashwa gandha)		Tonic for vitality
28.	*Aloe vera*		Anti-septic, cosmetics and many medicinal uses.
29.	*Stevia*		Sugar free, control of blood pressure.
30.	*Glycemai-gudmar*		Anti-diabetic
31.	*Acolius, Pathar chur*		Urinary and Kidney stone
32.	*Blue green algae*		Vitamin A,B,C,E. and Biotins
33.	*Guggul (Commiphora wightii)*		Lowering of cholestrol, anti-arthritis.

Drying

Most of the herbs are dried under shade. The drying under shade takes longer time but it is effective method of drying because it preserves aroma of herb to a large extent (b) nutritional content and (c) freshness that determine quality of herbs. The dried herbs are kept in air light packaging - The machinery required to manufacture herbal products are pulverizers, milling machines, cutting machines, sieves and bag packaging machines.

The herbs have a shelf life of 2-3 years. For export strict quality control is required and services of testing labs may be utilised for tests such as microbiological tests, _E. coli_ Tests, Salmonelle test, mould and yeast test.

High performance liquid chromatography is carried out to analyze ingredients of the mixture.

Gas chromatography (GC)- to separate each element to check purity of products.

Liquid Extract

For preparation of liquid extract steam distillation process is used:

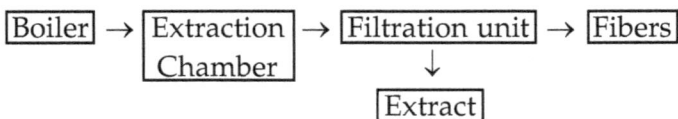

$$\boxed{\text{Boiler}} \rightarrow \boxed{\begin{array}{c}\text{Extraction}\\\text{Chamber}\end{array}} \rightarrow \boxed{\text{Filtration unit}} \rightarrow \boxed{\text{Fibers}}$$
$$\downarrow$$
$$\boxed{\text{Extract}}$$

The major equipment is

1. Baby Boiler
2. Extraction column
3. Filters
4. Collection Tanks

7
Herbal Sweeteners

There is growing incidence of diabetes all over the world and particularly in India it is spreading at a very fast pace and nearly 2.5 per cent population of the country is suffering from diabetes. Considerable research work is being carried out to Isolate diabetic molecule. Several herbal plants have become popular to control diabetes, like Gudmar, Jamun ki guthli, methi, Neem, Karela, Bijaisar etc. but very little attention has been given to identify plants which can be used as sweeteners without side effects. In America, Brazil, South Africa, Japan etc. herbal sweeteners have become very popular. It is high time we should start looking into development of sweetners like stevia, thaumatin, etc. In fact, there is National as well as International market.

Stevia is a pereniel shrub of Asteraceae family growing wild in Brazil and Paraguay. It is estimated that as many as 200 species of stevia are native to South America. But no other stevia plants have exhibited same intensity of sweetness as *S. rebaudiana*. It is grown commercially in many parts of Brazil, Paraguay Uruguay, Central America, Israel, Thailand and China etc. People of Brazil and Paraguay have used the leaves of stevia a sweetener in foods and tea. Stevia is considered hypoglycemic, hypotensive, diuretic, cardio tonic. The leaves are used for diabetes, obesity, cavities, hypertension, fatigue, depression etc.

Western interest in stevia began around in turn of nineteenth century. It was first studied in 1899 by Paraguay biotanist Moises S. Bertoni. Over 100 phytochernicals have been discovered. It is rich in terpenes, flavonoids, and glycosides. One of these glycosides in called stevioside which is sweetest and it is 300 times sweeter than sugar. The natural stevia leaf has been found to be non-toxic and has no side effects.

During the world war II, the allies considered extracting stevioside commercially as an alternative for sugar supplies which were running out. Unfortunately, at that time biotechnology was not available and there was restriction for the use of artifical sweeteners. In 1970 R and D in Japan made quick progress. For over 25 years the Japanese consumers have been using the extract from the plant as a safe, natural non caloric sweetner. It is currently the most used sweetener. The refined product stevioside is white crystalline powder extracted from stevia leaves. The use of stevioside sweetener has been cleared by U.S. Food and Drug Administration.

Agro Technology

In the wild, the plant grows in infertile, moist, sandy soil near streams and marshes. It reaches height of about 2 feet (3 feet by cultivation),with many branches and attractive, slightly serrated, opposite leaves, The pretty flowers are tiny and white with a pale purple throat, but these must be pinched off or these will steal sweeteners from the leaves.

Tropical and Sub-tropical climate suits cultivation of stevia. It can also be grown as a perennial crop in frost-free areas. Stevia should be planted in the field in the Jan.-Feb. when the temperature is about 65°F (18-20°C). A sandy loam soil with a slightly acidic to natural pH and good drain is appropriate. Stevia has a temperamental nature which is often reflected in slow growth but after a month, they pick up speed. As regards fertilizers using N:P:K::20:20:20 kg once a month during summer would be enough. During the summer, flowering takes place. Pinch off the flowers to produce sweetest leaves. Subsequently, when the fall starts it is time for harvesting the leaves. Pull up the plant in the early morning, remove the leaves and dry them in sun for about 24 hours or use solar driers for faster drying. When crispy dry, the leaves could be micro-pulverized and packed in airtight PVC Bags/containers.

One can set up nursery from harvested plants. Cut three inch pieces, dip in rooting hormone and insert in sand and mist frequently. Stevia cuttings can also be rooted in water which could be transplanted in Jan.- Feb.

Natural Sweetener Thaumatin

Thaumatin is a protein extracted from a West African indigenous shrub called *katemfe* and is locally used as a rural sweetener and flavour enhancer low-caloric extract, which is about 2,500 times sweet than sugar, has been cultivated and modified in the region for generations.

The protein itself was discovered by researchers at the Nigerian University. Although the katemfe shrub is indigenous to the forests of Western Africa and the Thaumatin protein was discovered by Nigerian scientists, intellectual property rights to the natural product lie in the US.

In the European Union, the natural sweeter is already approved for human consumption and is used in chewing gums, soups and other low caloric products. As the qualities of the katemfe-based sweetner are being improved; thaumatin is widely believed to become an important product. Currently, it is the most sold low-calorie alternative to sugar is Nutra Sweet, a big seller based on chemical compounds. As the healthiness of purely chemical products is often questioned in Western markets, a pure natural product such as Thaumatin is seen as a clear market winner. Alone the US market for low calorie sweetners is set at US$ 900 million annually.

Biotechnological Approach

In the US, researchers from the University of California and the Lucky Biotech Corporation have received a patent for all transgenic fruits, seeds, and vegetables containing the gene responsible for producing Thuamatin. The multinational food company Unilever has already made successful attempts to insert these genes into bacteria. Thaumatin production based on genetic manipulation of bacteria would put an end to all Ghanaian revenues from the katemfe shrub. According to Genetic Resources Action International (GRAIN), "the genetically engineered route for the production of thaumatin is far cheaper than harvesting it."

Thaumatin is a sweet-tasting protein extracted from the arils of the fruit of *Thaumatococcus daniellii*. This plant is indigenous to Africa. Its fruits have been used for centuries to improve the taste of the locally produced palm wine, and as a sweetener in cooking. Aqueous solutions of Thaumatin have an intense sweetness, about 5,500 times that of sucrose when compared with a 0.6 per cent sucrose solution, and about 1,600 times in relation to a 10 per cent sucrose solution. For Western and Oriental markets, it has been used as an intense sweetener and as flavour potentiator capable of improving or extending the flavour charcteristics of various food products. Being soluble in aqueous ethanol, it can be added directly to flavour essences or oil. In Japan, it is used in a range of food products but in other countries it is only used in chewing gum and pharmaceuticals. Taste enhancing effects of thaumatin were observed just below the threshold concentration (5x10 per cent) of thaumatin. Tallin (the trade name for thaumatin produced by Tata and Lyle (PLC) not only enhances flavours associated with sweet products such as peppermint but is also capable of improving savoury flavours. It reduces the harshness of tobacco and improves its 'smoothing' effect when incorporated in filter tips. A flavour-enhancing effect in cultured milk has also been reported.

Various types of sweeteners are available in market such as synthetic sweetener from aspatame. Sugar free natural extracts from sugar and stevia. CFTRI Mysore has standardised the process of stevia powder. Preparation of liquid extract from leaves, vaccum shelf drying at low temperature is used or spray drying could also be dose.

FOOD PROCESSING

8
Preservation of Fruits and Vegetables

A large variety of fruits and vegetables grow under sub-tropical and temperate climatic conditions due to diverse agro-climatic conditions prevailing in different regions of the country. India produces annually 32 million tonnes of fruits which is 12 per cent of worlds' production and 71 million tonnes of vegetables which forms 15 per cent of world production. The major fruits grown are banana, mango, citrus (orange, mausami), guava, grapes, apple, pineapple etc. Although India is a very large producer of fruits per capita consumption is mere 75 gm. Further, 20 to 22 per cent of production of fruits is lost due to spoilage at various post harvest stages and transportation which amounts to Rs.3000 crores in a year. Nearly 76 per cent of production of fruits and vegetables is consumed fresh while

Micro-Enterprises in Agriculture

losses account for 20-22 per cent. Only 2 per cent of vegetable production and 4 per cent of fruit Production is being processed compared to Brazil-70 per cent, Malaysia-80 per cent, Philippines-78 per cent, and Thailand-30 per cent There are 7000 Processing units in organized sector and thousands in unorganized sector and the industry employs 1.6 million workers. The turnover of industry is Rs.3,00,000-3,50,000 crores. The demand for processed fruits and vegetables comes from both domestic and export markets. A substantial share of production is consumed by the defence department, Hotels and restaurants. Household consumption accounts for less than 50 per cent production. Export of processed products is 13 per cent of total exports of the country. The bulk share in exports is of mango products. India's share in world trade in this sector is around one-two per cent.

The average Indian farmer is poor, high quality planting material is not easily available and he is not fully aware of modern agro economic practices. The yield per acre is quite poor compared to other Asian countries.

The storage and transport infrastructure is highly inadequate resulting in high post-harvest wastage and losses. The average Indian farmer is hesitant to take up horticulture requiring high investment, long gestation and uncertain returns.

The major constraints for processors are:

1. Non-availability of high/uniform quality of fruits and vegetables.

2. The domestic demand of processed foods is low due to availability of fresh fruits.

3. Absence of captive farming and corporate farming.

Constraints of consumer is high cost of processed foods, which should be brought down to suit the pocket of middle class consumer.

Fruit and Vegetables Preservation/Process Flow Sheet

Processing of Range, Lemon and Mango

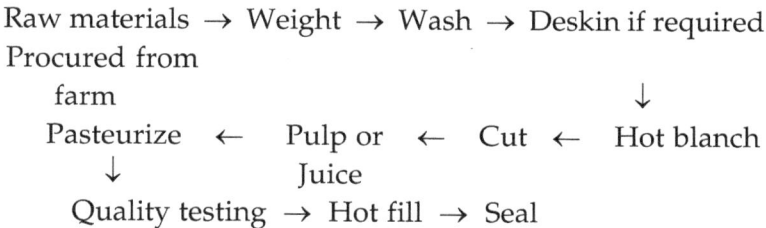

Raw materials → Weight → Wash → Deskin if required
Procured from
    farm   ↓
    Pasteurize ← Pulp or ← Cut ← Hot blanch
      ↓   Juice
    Quality testing → Hot fill → Seal

One can process all types of fruits and vegetable using the same machinery with minor addition/alteration. Thus a wide product mix is obtained. One particular fruit is processed during certain duration when raw material is available at cheap price or there in bumper crop.

The plant could be setup in rural or semi-urban areas to economise the cost of transportation. A direct linkage with farmers is suggested for easy and cheap rates of raw materials.

Central Food Technological Research Institute, Mysore (CFTRI) has developed few hundred processes/products to cover wide spectrum of raw materials ranging from cereals, pulses, grains, spices, fruits and vegetables, oil seeds to animal products. Entrepreneurs should approach them for Technology and Training etc.

In addition to normal means of finance, Ministry of Food Processing Industries (MFPI), offers promotional incentives like soft loan, subsidies etc. The entrepreneurs should contact the Ministry to avail their assistance. For marketing and exports "Agricultural Processed Food Products Export Development Authority (APEDA) New Delhi should be contacted.

All the machinery like pulpers steam Jacketted vessels, vacuum shelf driers, spray driers, hot air driers are available indigenously.

9
Cryogrinding of Spices/Herbs

The science of cryogenics belongs to very low temperature in the range of -150°C to -238°C. The absolute temperature achievable is -273°C- 0°K. The materials subjected to above temperature range (-150°C-238°C) become very brittle and can be broken to pieces with little impact. Their behaviour change at very low temperatures. During world war-II it was observed that materials frozen (at low temperature) are more resistant to wear.

Science of cryogenics has a wide range of applications from super conductivity (loss less flow of electricity), cryogenic engines, cryosurgery, cryogrinding etc.

Spices are most important part of Indian life. These are used in Indian foods and cuisines in every house hold, restaurant, hotels etc. Commonly used spices and cuisine are turmeric, chilly, black and white pepper, ginger, garlic, coriander, cinnamon etc.

Cryogrinding is new concept in grinding by which one can do micro-grinding to smallest particle size which is difficult to achieve by conventional grinding at ambient temperature. In cryogrinding particle size reduction is, carried out under inert atmosphere and reduces emission of volatile part. The grinding is done under Nitrogen atmosphere.

It improves aroma and minimize loss of essential oils - 3-10 per cent compared to 15-43 per cent in conventional grinding. Moreover, particle size is reduced upto 50 micron compared to 500-1000 micron in conventional process. Life of equipment is enhanced, avoids heating and colour of species is retained.

India is largest exporter of spices amounting to US $ 500 million (335488T) according to 2005-006 figures.

Table 9.1: Share of Exports

India	25 per cent
China	24 per cent
Spain	17 per cent
Mexico	8 per cent
Pakistan	7.2 per cent
Morocco	7 per cent
Turkey	4.5 per cent

In order to retain leadership in export of spices there is need to adopt new technologies for improving quality.

Equipment required is such as liquid nitrogen supply, blower, rotary filter, impact mill, recovery system for N_2 etc. cost of machine is Rs. 120-150 lakhs.

There is suggestion to carry out grinding in cold chamber by using common refrigeration system. It will be at least better than conventional grinding. CSIR has developed mustard oil expeller with water jacketted system to preserve Aroma of Mustered oil and it is very successful. Similar system could be used for spices.

10
Convenient Food–
Gravy Mix

Now, it is an era of convenient foods and convenient foods under various brands are becoming popular. Now a days, husband and wife are both working professionals and hardly get time to prepare food and even they don't know how to prepare all items. They reach home late and bring packed food stuffs to home. In any Indian Kitchen substantial time is required to prepare masala by grinding onions, garlic, ginger and make paste after mixing with spices. This paste is then fried in cooking oil/ghee for 5-10 minutes till it becomes brown. Water is added to the fried paste. The whole process takes 15-20 minutes. Gravy is a convenient alternative to this drudgery. The gravy in the form of powder is added to cooking oil in cooker with water and it takes 30 seconds, finally cut vegetables etc. are added to curry, and your dish is ready in 10 minutes. There are more than dozen formulations (developed by CFTRI Mysore) to suit a variety of veg and non veg dishes.

The equipment required are peelers, slicers, hot air drier vacuum shelf dryer, SS grinder, weighing and packing machine. Investment of Rs. 20 lakh on plant machinery is required. One can prepare 30 tonnes of gravy mix and pack in metallised polyester pouches.

11

Iodized Salt

All human beings including animals require adequate quantities of Iodine in food to make the normal amount of thyrosine through thyroid glands. Due to deficiency of Iodine the glands enlarged and form growth on neck which is known as Goitre. Intake of iodine is also necessary for pregnant and lactating mothers and children for healthy growth. The Iodized salt is the best mean of replenishing iodine deficiency. Pregnant and lactating mothers require 175-200 mg of Iodine daily.

In India large manufacturing companies have captured more than 80 per cent, but it is beyond the reach of poor people because of very high cost market. Small entrepreneurs should come forward for setting up micro-enterprises to meet the needs of weaker section of society.

There are various methods of Iodization developed by CSIR Central Salt and Marine Chemical Research Institute

(CSMCRI) Bhavnagar such as continuous roller process with spraying arrangement, batch process and immersion process, the last is most suitable for small entrepreneurs.

In the immersion process, a concrete tank is constructed for preparing mix of Potassium Iodate (KIO_3) in water. 30 gm of KIO_3 is dissolved in the tank for processing one ton of raw salt. Raw salt is filled in buckets/baskets and dipped in Iodate solution for 5 minutes and then taken out. It is then spread over a platform for drying in the shade or in the night. The dried raw salt is then ground using roller grinder. The process is very simple.

Suggest capacity of the unit is 02 Ton/day.		
Land and Building	100 sq.m.	2.00 Lakhs
Civil work Tank and platform for drying		0.50 lakhs
Roller grinder 200 kg/hour 3 h.p. (3 phase)		0.10
Weighing machine		0.10
Packaging material		0.10
Misc.		1.00
Total		**3.80**
Working capital (45 days)		**2.00**

At the processing level presence of Iodine is 30ppm which is reduced during grinding and drying. At any time Iodine content should not be less than 15ppm, when in use.

Entrepreneurs should contact CSIR (CSMCRI Bhavnagar, Gujarat and also contact Salt Commissioner (GOI), Jaipur to get permission for setting up unit.

12
Energy Food

Malnutrition is one of the major problems of the country due to lack of appropriate diet among the weaker section of society, lower middle class as well as growing up children. Lack of protein and calorie impairs the physical and mental growth of children and pregnant women generally suffer from mal-nutrition. Central and State Governments are very concerned to combat these deficiency disorders and have launched several nutritional intervention programmes. Child and woman development deptt., tribal welfare and rural development deptt. operate nutritional deficiency programmes to replenish the protein/ calorie requirement.

There are variety of nutritional foods available in the market but the price of these foods is exorbitant and beyond the reach of middle class population. There is need to introduce cheaper supplementary foods within the reach of common man.

CSIR's Central Food Technological Research Institute Mysore, have introduced few formulations based on wheat, Bengal gram dhal, Soya/ground nut-flour Jaggery with vitamins and minerals and these are distributed under various healthcare programmes and these have proved very useful.

Author has modified this formula and has used chocolate powder, milk powder, cardamom, sugar instead of Jaggery for longer shelf life and better taste. One or two spoons of the modified mix could be added in water on milk.

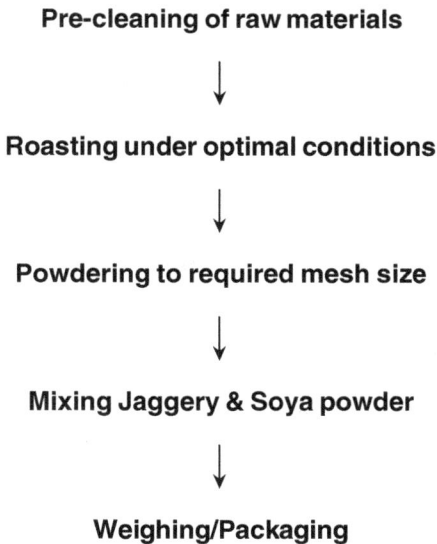

Pre-cleaning of raw materials

↓

Roasting under optimal conditions

↓

Powdering to required mesh size

↓

Mixing Jaggery & Soya powder

↓

Weighing/Packaging

Process Flow Sheet

The grains are first of all cleaned to remove dust and stones etc. The cleaned ingredients are roasted to optimal conditions such to develop golden colour and pleasant aroma of wheat and then bengal gram is roasted till it gets a fine aroma. The roasted grains are cooled cleaned and again destoned. Subsequently, the roasted grains are ground in hammer mill. Similarly, Soya flour and defatted groundnut are roasted in electric roasters, cooled and ground to desired mesh size. Jaggery is broken into small bits, mixed with calcium carbonate and some wheat flour and passed through multi-mill to get course flour.

The mix has pleasant flavour and good acceptability and could be consumed as such and requires no elaborate cooking as in the case of other available high protein supplements. If desired it could be mixed with water or milk to make porridge, paste etc.

Coco powder or elaichi, milk powder, sugar powder, may also be added to roasted grain/Soya powder to make super formulation for upper strata of population.

The products could be introduced directly in market supported by cheaper local TV ads, or recommended by medical practitioners, NGOs in health care, Mahila and Bal Vikas Deptt. etc.

Equipment/machinery include.

1. Cleaners/Destoners
2. Roasters
3. Hammer will
4. Mixer
5. Weighing and packing machine.

One can start with 1-2 tonn/day capacity with investment with Rs. 25.00 lakhs

Technology is available through CFTRI Mysore.

BAKERY

13
Biscuits

Biscuits have more food value with substantial energy, protein, carbohydrates and minerals apart from the good taste. Further, vitamins can be incorporated in the biscuit's recipe that are essential for balanced diet at a reasonably low cost. Consumption of biscuits is gaining more and more popularity. There are at present about 60,000 bakery units in the country. The leading producers are the DGTD registered units numbering 54. As many as 35 of these are registered for bistuits production and only 19 for bread production. The number of large units is not more than 22. The medium sized units are estimated to be 400 in number. The number of small units are 4000 while the household units are the largest in strength at 56,000. It is likely to rise output to 35,00,000 tonnes valued at Rs. 8,000-10,000 crores. The demand of this group of biscuts will increase with population.

There is big demand for Atta and multigrain biscuits, cookies, Khataies etc. Also there is need to set up cottage scale units to produce energy biscuits using Soya flour (having 40 per cent protein).

Raw Material

The major raw materials required are maida, atta-grains, soyaflour, coconut powder, salt and vanaspathi (fat) and the miscellaneous items are ammonium bicarbonate, baking soda, essences and milk powder. All the ingredients are available locally.

Plant and Machinery

☆ Grinder

☆ Flour sifter

☆ Sigma dough mixing machine

☆ Moulding machine

☆ Baking oven/oil/gas fired/electric

☆ Oil sprayer

☆ Packing tables

☆ Auxiliaries like–Heat sealers trays, exhaust fans and office equipment

Capacity of the unit	: 300kg/day
Working days	: 300
Optimum capacity utilization	: 70-80 per cent
Cost of machinery	: Rs. 5.00 lakh
Working capital per month	: Rs. 4.0 lakh

Source of Technology

CFTRI has standardized the technology and general method of processing of different types of biscuits. However, the formulation/recipe can vary to make different type of biscuits within the level of quality standards. Apart from this procedure for quality control, packing and packaging material specifications, equipment details are also provided by the institute.

The manufacturers have to take a license under PFA (Government of India).

14
Shelf-Stable and Ready-to-Eat Foods

Introduction

Recent development in plastic materials and thermal processing technologies have made it possible for the many lip-smacking Indian dishes to be available in ready-to-eat form. Vegetable Biryani, Navarathan Korma, Bisibele Bhath, Palak Paneer, Sweet Pongal-you name any Indian ethnic dish and you have it literally on the platter. There is no need to depend on the grandma or a gourmet chef. All that one has to do is to pick up the product packed in retort pouch or tray, heat in boiling water for 5 minutes, cut open the pack and serve. One can as well use a microwave oven instead of boiling it in water.

Market Potential

These "heat and eat" products own their success to the ultimate convenience, they offer to the consumer. They are shelf stable, which means they can be stored at room temperature without requiring refrigeration and in their packed form, remain fresh for over one year. And all this without any added preservatives. Singles' house-holds, on-the-run professionals, double income families and harried housewives form an instant market for these products, especially in the metros. Besides, there are huge export markets. Some products are already in market 'Sarso ka Saag' and Dal Makhni are exported to Europe and USA from India.

Advantages

Retort pouches and trays offer many advantages compared to canned and frozen foods. The pouch/tray with its thin cross-sectional profile compared to can require about 25-50 per cent less process time. The rates of destruction of sensory and nutritional factors being greater than that of microorganisms at lower temperatures, quick heat penetration in pouches/trays helps in better quality retention. Because of the shorter process time, many of the delicate Indian dishes can be processed without losing colour, flavour and texture.

Technology has been developed by CFTRI for a variety of ready-to-eat Indian foods-both vegetarian and non-vegetarian products that include Vegetale Pulao, Tomato Rice, Puliogera (Tamarind Rice), Bisibele Bhath (Sambar Rice), Plain Rice, Ghee Rice, Vegetable Korma, Aloo Matter,

Channa Masala, Dal Fry, Methi Dal, Palak Paneer, Navarathan Kurma, Shahi Paneer, Shahi Rajma, Sambar, Sweet Pongal, Coconut Chutney, Chicken Biryani, Chicken Curry and Mutton chops.

Process Description

Product Preparation → Pouch Filling → Pouch Sealing → Visual Inspection → Racking and Retort Loading → Retort Unloading → Drying of Pouches → Cartoning and Casing.

Raw Materials

Cereal grains, Pulses and legumes, Green leafy vegetables, Roots and tubers, Fruits, Nuts and Oil seeds, Condiments and species, Milk and milk products, Fats/Oils, Sugars, Salt and Saffron.

Equipments

Principal Equipments

Heat sealer, Compressor for heat sealer, Retort, Compressor for retort operation, Boiler, Pulveriser, Peeler, Slicer, Steriliser, Grinder, Steam Jacketted kettles, Racking system, Kettles with stirrer, Frying system, Centrifugal pump, LPG stove with gas connection, Pre-heater, Liquid filling machine etc.

Auxiliary Equipments

Deep Freezer, Walk in coolers, Balances (Table top and Digital), Hydraulic pallet truck, Generator, Material handling equipments, working tables, storage racks and Laboratory equipments etc.

Techno-Economics

Suggested economic capacity: 1000 kg/day or 300 T/anum (for four products)

Working: 1shift/day, 300 working days/year.

Optimum utilization capacity: 70 per cent

Land 1600 sqm. Building 800 sq. m.

Machinery: Rs. 75-80 lakh

15
Dairy Industry

There is acute shortage of milk and milk products in urban areas inspite of "white revoluiton" more than two decades back. The problem becomes aggravated during the summer season when milch cattle start drying up due to heat. In tropical and semi tropical countries many difficulties arise during handling of fresh milk, since the milk remains sweet for only short time during summer. It is further required to be transported to long distances. High bacterial contamination of the milk leads to physical and chemical changes which makes it unsuitable for consumption or for conversion into milk products. The keeping quality of milk is enhanced by artificial cooling on the farm and during transportation. The delay in cooling result in higher bacterial count of milk with consequent detoriation in its quality. Milk is therefore chilled to 4°C soon after milking, so as to suppress the bacterial growth and it is kept at 4°C till it is subjected to further processing at the dairy plant.

Various methods are used for chilling the milk such as keeping the milk in double walled vessels using ice as cooling media, keeping in tank surrounded by ice and salt (brine), a tank in which cooling media (refrigerator) or chilled water flows through tubes of the cooler while milk flows in a thin film our the exterior of the S.S. tubes etc.Plate heat exchanger which consist of a number of S.S. plates assembled about 0.69 cm apart. The process being carried out in a closed system. The milk is not exposed and could be cooled to 4-6°C. Plate - heat exchangers are very compact and occupy must less space than tubular surface cooler of same capacity.

The chilled milk is stored in insulated tanks. The insulating malerials used are cork, onazote, Thermocole or glass wool. If tank is properly insulated the rise in temperature should not be more than 01°C in 12 hours. Refrigerated tanks are also used for final chilling of milk and storage depends upon the availablty of electricity - The refrigerated tanks are more costly.

The next step is clarification *i.e.* removal of any sediment using centrifugal clarifier. During clarificaton apart from removal of sediments also some bacteria present in the milk is removed.

The homogenisation is required to subdivide fat globules with smaller globules so as to cause the milk to lose its creamy property. The process increases the viscosity and slightly richer milk. The milk is heated to 60°C so as to inactivate lipase and then pumped through an orifice at a very high pressure.

It may be necessary to manufacture milk of standard composition such as in case of toned milk. The fat content of standard and whole milk should be 5.0 per cent butter fat and toned milk contains 0.1 per cent fat.

Pasteurisation is very important to remove disease providing organism present in milk. There are various methods of pasteurisation. In Holder method milk is heated to 62°C and kept at this temperature for 30 minutes and is cooled below 10°C. The pasteuriser is a double jacketted vessel. The milk is agitated continously and steam passed through the tubes fixed between annular space. The method is suitable for 900-1000 L/hour capacity.

In continous process a vessel is used having 4 compartments. Milk is heated to 62-68°C in a pre-heater in a batch pasteuriser and kept for 30 minutes and consequently flows through remaining compartments.

The entire process is continuous, being mechanicaly controlled with timers. Cooling is done with a surface type or plate type cooler. The process is suitable for small scale capaicity of 700 litres hr and above.

In High Temperature Short time (HTST) method milk is heated to 71°-72°C for less then 15 seconds and immediately cooled to temperature below 10°C. The entire process is continous and automatic and electrically/pneumatically controlled. Temperature should not exceed 72°C to maintain creaming quality of milk.

Before operating the plant it is sterlized by circulating water at 85°C for less then 10 minutes. If necessry good quality detergent could be used to wash the plant but remains of detergent and residul water should be extricated.

In uperisation process, the milk is allowed to flow and heat is applied by direct injection of steam. The milk is heated upto 80°C through tubular pre heater and deaerated before it reaches uperisation tube. The milk is then heated to 80°C for a second by direct injection of high pressure steam to the uprisation tube. This is followed by instant cooling to 60°C under vacuum. The injected steam is removed with unpleasant odour present in steam. The milk is then transferred to a cooler and finally stored in cold room. CSIR, Central Scientific Instruments Organisation, Chandigarh has also developed a novel process for pasteurisation but it is more costly. The chilled milk at room temperature is passed on to asceptic packaging machine for sealing in PVC pouches.

The enterpreneures can choose the machine according to investment involved. They can further set up solar water heating system for pre-heating water to 70-80°C for feeding to boilers to save in cost of fuel etc. The micro-level or cottage scale entreprenures can set up machinery equipment for down stream products like cream for supply to ice cream industry, production of Ghee/Curd, Butter, Paneer, Khoa, Cheese, Srikhand. For training and technology one can contact National Dairy Research Institute (NDRI) Karnal, Haryana. The machines are available for producing down stream products.

16
Natural Vinegar Production

The word "vinegar" is derived from the Old French *vin aigre*, meaning "sour wine". Vinegar has been used since ancient times and is an important element in European, Asian, and other traditional cuisines of the world.

Vinegar is an acidic liquid processed from the fermentation of ethanol in a process that yields its key ingredient, acetic acid (also called ethanoic acid). It may also come in a diluted form. The acetic acid concentration typically ranges from 4 to 8 per cent by volume (typically 5 per cent) and higher concentrations for pickling (up to 18 per cent). Natural vinegars also contain small amounts of tartaric acid, citric acid, and other acids.

Vinegar production is a traditional rural industry and there is scope to improve the process of vinegar production.

Natural vinegar can be prepared from raw materials like sugarcane or black berry (*Jamun*). These materials are available in abundance in rural areas. Vinegar is used for sprinkling over salad, pickles and many food preparations. There is also a market for vinegar in cities. It is a profitable rural industry.

History

In ancient times vinegar was produced from sugarcane juice and fruits containing sugar or starch, wastes like fruit peels, carrot peelings, etc, by utilizing bacteria and yeast in air to colonize the juice leading to fermentation. The process included sterilizing the container, keeping for 5 minutes and pouring raw material in the container and covering with cloth. The material was kept for few weeks before opening the cloth. The vinegar was ready when its odour could be smelt. The liquid was then filtered.

In the improved process, the fruit juice is poured in the sterilized container and 50-100 ml of natural vinegar is added for culture in 10 litres. The container is then covered with cloth or plastic lid and thereafter covered with thick jute or woollen cloth so that approximately 35°C temperature is maintained for 3-4 weeks. The lid is opened and the odour of ready vinegar emits out. The liquid is filtered and sterilized for 5 minutes at 50°C. It may be noted that distilled water should be used to dilute fruit juice and container is filled to three by four (3/4). In this vinegar acetic acid (CH_3COOH) content should be 4-5 per cent. The finished product is packed in glass bottles.

For making table vinegar used in hotels, etc, 4-5 per cent pure acetic acid is mixed with 95 per cent distilled water. It is white in colour or sometimes colour is added. Continuous use of this synthetic vinegar may harm the intestines, etc.

Raw Material for Natural Vinegar Production

Sugarcane juice and black berry are commonly used. We can use sweet fruit juice like sapota, apple, raspberry, strawberry, carrot, etc. The natural vinegar find use for removal of dandruff, cough/cold, headache, vomiting, upset stomach or indigestion, to reduce cholesterol, antiseptic, cleanser, etc. The vinegar is also used for sprinkling over salads, sauces, chutneys, etc.

For eradication of dandruff wash hair with shampoo and apply apple cider vinegar regularly for some period till dandruff vanishes. Take two part of vinegar plus one part of water and a few drops of lemon grass oil or rosemary oil or lavender oil, apply the mix to the hair.

Black Berry Vinegar

About Black Berry

Jamun is a very common, large evergreen beautiful tree of Indian subcontinent. The scientific name of *Jamun* is *Eugenia jambolana* or *Syzygium cumini L* and it belongs to the myrtaceae plant family. Common names are java plum, black plum, jambul and Indian blackberry. It grows naturally in clayey loam soil in tropical as well as sub-tropical zones. It is widely cultivated in north as well as the rest of the Indo-Gangetic plains on a large scale. Its habitat starts from Myanmar and extends up to Afghanistan. It is generally cultivated as a roadside avenue tree as well.

Properties of Black Berry

Jamun vinegar is not only an appetizer but also helps indigestion. *Jamun* in Ayurvedic terms is an astringent and sweet fruit. Vinegar is made from the juice of slightly unripe, tangy purple *jamuns*, which are otherwise eaten with a dash of salt. The ripe *jamun* is a carminative, digestive, coolant and liver stimulant and jamun vinegar has similar properties. *Jamun* is favoured for its diabetes controlling powers. Tests at the Central Drug Research Institute, Lucknow, indicate that ingesting the alcoholic extract of the seeds reduces the level of blood sugar and glycosuria.

Uses of Black Berry

☆ This vinegar is an agreeable stomachic and carminative.

☆ It is used in the treatments of various diseases such as stone, diabetes,

☆ constipation and jaundice, the *jamun* vinegar is highly in demand.

☆ *Jamun* fruits are a general tonic, cooling and astringent to bowels.

☆ These enrich blood and strengthen liver.

☆ It can be used in various cuisines, mainly Italian and Indian.

☆ Ideal for salads, marinades, sauce vinaigrette, French/Italian dressings.

☆ It can be used in various meat and seafood dishes and also for preserving pickles and pastes.

About Sugarcane

Sugarcane, is native to warm temperate to tropical regions of Asia, they have stout, jointed, fibrous stalks that are rich in sugar and measure 2 to 6 meters (6 to 19 feet) tall. All sugarcane species interbreed, and the major commercial cultivars are complex hybrids. About 195 countries grow the crop to produce 1,324.6 million tons

Sugarcane Vinegar

(more than six times the amount of sugar beet produced). As of the year 2005, the world's largest producer of sugarcane by far is Brazil followed by India.

Properties of Sugarcane

Owing to all culinary properties, it is highly recommended in indigestion and acidity problems. Being manufactured from 100 per cent natural sugarcane, it has no colour or is colourless. The essential aspect of this vinegar is that it has no side effects on the body.

Uses of Sugarcane Vinegar

Sugarcane vinegar is used for the treatment of diseases like constipation, stone and jaundice.

☆ It is a good taste maker

☆ It prevents obesity if taken with sufficient water

☆ No acidity or indigestion if occurrs used in food regularly

Micro-Enterprises in Agriculture

☆ Very useful for summer heat boils

☆ It improves health, no side effect

Process of Vinegar Production

The process involves keeping the pulp and juice (of *Jamun* or Sugarcane) in wooden vats or large earthen pots with the addition of culture. The pots are kept for 15-20 days duly covered with Jute cloth to provide heat and accelerate fermentation. Thereafter, the material is filtered and sterilized by application of slow heat. The finished product is packed in glass bottles.

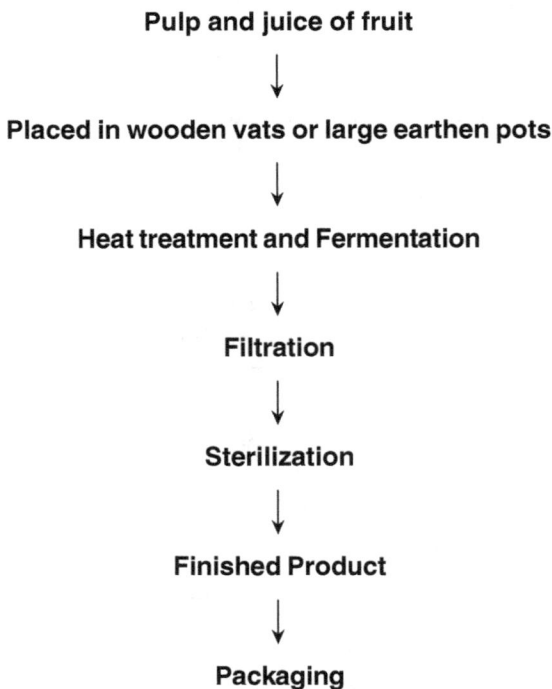

Pulp and juice of fruit

↓

Placed in wooden vats or large earthen pots

↓

Heat treatment and Fermentation

↓

Filtration

↓

Sterilization

↓

Finished Product

↓

Packaging

Flow Chart of Process of Vinegar Production

Economics

1.	Capacity	5000 litre/year
2.	Vats and utensils	Rs. 1 lakh
3.	Working capital	Rs. 1 lakh
4.	Total investment	Rs. 2 lakh
5.	Sales 5000 L x Rs. 50/- per litre. (ex-factory price)	Rs. 2.50 lakhs
6.	Cost of production (including packaging)	1.00 lakh
7.	Profit	Rs. 1.50 lakh

Source of Technology

CFTRI, Mysore and Traditional Industries.

17
Spirulina

Spirulina is a simple, single-cell form of blue-green algae that gets its name from its spiral shape. This simple organism has been around for millions of years, but is currently being hailed as the superfood of the future because of its exceptional nutritional content. Spirulina is a better source of protein than either beef or soybean. It is also one of the few non-animal sources of Vitamin B12, and contains twice the amount of B12 found in beef liver, which makes it an excellent addition to the vegetarian diet.

Spirulina also contains vitamins A, B, C and E, as well as health-enhancing amino acids; significant amounts of the gamma-linoleic acid (GLA), a fatty acid that promotes cardiovascular health; chelated forms of the minerals potassium, calcium, magnesium, zinc, selenium, phosphorous, and iron; complex sugars, trace elements, and enzymes–all of which are easily absorbed from spirulina by the body. It also contains chlorophyll (green) and phycocyanin (blue) pigments in its cellular structure, which give it a blue-green colour, it has been proven that it helps to detoxify the liver.

Spirulina is best known for its ability to boost the immune system, and research indicates it may even help both treat and prevent cancer. Spirulina stimulates the natural killer (NK) cells that fight illness and attack and kill cancerous cells. In one study performed in India, participants taking spirulina saw a complete remission of mouth cancer.

Egyptians used Spirulina as a food supplement 5000 years ago obtained from natural resources. But its cultivation started in Mexico in 1640. However, it received impetus recently (after 1950) due to its use as space food.

It has highest protein content in the range of 61-65 per cent. In any natural produce one gram of spirulina is equal to 1 kg of fruits and vegetables in terms of nutrition content. It is used for slimming without loss of energy in western world. Use of spirulina for eradiction of malnutrition among BPL families is well-known as some NGOs are operating in such programs. Small scale industries in countries like Mayanmar and Thailand are exporting spirulina products.

The price of spirulina in powder form is about Rs 1000/ - per kg while it is Rs 2000/- per kg in capsule form.

In India, several companies are producing spirulina using indigenous technology. Some brand names have become very popular for following applications:

Spirulina Tablets

☆ Treatment of anaemia

☆ Control of diabetes

☆ Lowering of cholesterol

☆ Healing of wounds

☆ Food colourant

☆ Protein supplement

Medicinal Value of Spirulina

Spirulina supplements are available in health food stores and from online distributorships in bulk powder, capsule, and tablet forms. You can also find it as an additive to the many new "green drinks" available in the market. People who are vegetarians or who have a poor diet or one loaded with processed food should definitely consider taking this supplement–it is a safe, non-toxic, and very efficient way to meet your total nutritional needs. In addition, many people reported that adding spirulina to their diet gave them extra energy and in some cases eliminated feelings of fatigue.

Reported Uses

Spirulina has been used in diet and weight-loss products for its high nutritional value and claimed action on appetite suppression. There are reports of its use instead of dietary supplements, but its cost does not justify its use in this manner. In developing countries, such as Peru, India, Vietnam, and Togo and other African countries, spirulina is used to help fight protein and vitamin A malnutrition. In industrialized countries, the GLA content is thought to contribute to the prevention of cardiovascular (CV) disease.

A double-blind, placebo-controlled, cross-over study was conducted to evaluate spirulina's effect on weight reduction. Sixteen patients already enrolled in an outpatient dietary self-help group took part in this 4-week trial. Patients were asked to ingest 14 spirulina tablets (Verum: spirulina 200 mg + synthetic vanilla) or placebo (spinach powder 200 mg + synthetic vanilla) immediately before each meal three times daily. Patients were evaluated for changes in body weight, biochemical variables, blood pressure, heart rate, and adverse effects of treatment (by questionnaire) at 2-week intervals. Each treatment phase lasted for 4 weeks with a 2-week washout between phases. At the end of the study, the spirulina group had dropped an average of 1.4 kg in weight, whereas the placebo group had dropped an average of 0.7 kg. The difference between the two groups was not statistically significant, but the investigators suggested that the results were sufficiently promising to warrant pursuit of a longer-term trial. Concerns exist with respect to the trial's study design (small sample size, unclear blinding and randomization techniques) and short duration.

Calcium spirulina (Ca-SP), a polysaccharide derived from spirulina, has demonstrated inhibition of replicating viral cells (similar to anti-retroviral mechanistic activity) *in vitro*. An inhibition of heparin cofactor II-dependent antithrombin activities has been shown *in vitro* as well.

Simultaneous treatment with Ca-SP and tissue plasminogen activator (TPA) results in a synergistic enhancement of TPA production.

Other reported uses for spirulina include treatment of anaemia, diabetes, glaucoma, hair loss, hepatic disease, peptic ulcers, pancreatitis, and stress.

The process involves inoculation of spirulina culture in pucca tanks having mechanised agitators to oxygenate the water. About 20-25 gm of Spirulina grows in 1.0 sq. meter surface area of water in a day. The blue-green algae is removed from water surface and allowed to dry before purification and production of powder by spray drying process.

For production capacity of 6 Tonne/year of Spirulina powder, there is fixed capital investment of Rs. 40.00 lakhs which includes construction of water tank of 2000 Sq. m. and equipment like filterig unit, spray drier, vacuum packaging unit and packaging material. It is suggested that farmers may construct tank of size 20mx20m and grow algae under hygienic conditions and supply the green algae to mother plant having refining and spray drying facilities. This will involve large number of farmers and reduce individual investment and enable the plant to achieve higher producion capacity. The mother plant could supply powder to pharmaceutical industry for packaging in the form of capsules/sugar coated tablets.

Process

The tanks are filled with fresh water and spirulina culture is inoculated over the water surface. After thick layer of blue-green algae is formed over the water surface it is removed and allowed to dry over *pucca* platforms before supply to mother centre (lab) for purification and production of spray dried powder for supply to pharma industry.

Techno-economics

(*a*) Capacity – 6 tonne/year

(*b*) Construction of tank 2000 sq m.

☆ Filling unit

☆ Spray drier

☆ Vacuum packaging equipment

☆ Misc. lab apparatus

Rs.40.0 lakh

Spirulina Cultivation

Tanks filled with fresh water

↓

Spirulina culture is inoculated

↓

A layer of blue-green algae is formed

↓

A layer is removed and allowed to dry

↓

Supplied to mother culture (lab) for purification and production of powder

Process Flow Chart

The capacity could be enhanced for higher profits if large number of farmers could be involved. Entrepreneurs/ NGOs could also approach organisations like Department of Biotechnology, New Delhi, for soft loans and other benefits.

Source of Technology

☆ Director, CFTRI Mysore

☆ Director AMM Muragappa Chettier Research Centre, Taramani, Chennai

☆ NGOs working in this field.

18
Sugarcane Juice–Beverage

Sugarcane juice is a very popular drink in India and other Asian countries. At present it is being produced by vendors using hand operated or motorized crushers and served within a short time. The Juice cannot be stored even for a couple of hrs due to initiation of fermentation process. The juice is produced and served under very unhygienic conditions and hygiene conscious public does not opt for sugarcane juice. Hygienically produced and bottled sugarcane juice could find use as other soft drinks and has good potential market as the juice is nutritious containing minerals like iron,

magnesium, phosphorous and calcium and organic acids like malic acid, succinic acid and acetonic acid. In Ayurvedic system of medicine, sugarcane juice is recommended for use by jaundice patients.

In the process of manufacture first step is to clean sugarcane by soaking in water for an hour, scrubbing with nylon-coir brushes for cleaning the sugarcane and then dipping in 1 per cent potassium metabisulphite (KMS) solution for 5 minutes. The cleaned canes are then crushed in pre-cleaned crushers, filtered through filter cloth and collected in SS containers (rinsed with 1 per cent K.M.S.) Preservatives are added to the juice which is finally filtered through filter press. The juice is diluted to 15 Brix by using deionised water. The juice is then blended with fruit juices and is filled in 200 ml bottles, crowned and then pasteurized at 70°C for 10 minutes. The bottles are now cooled and stored for supply.

Investment of Rs. 30.00 lakhs is required for a capacity of 6000 bottles per day. Technology is available from CFTRI, Mysore or Agro Processing Division, Tamil Nadu Agriculture University, Coimbatore.

Techno-Economics

Duration of work	225 days
Capacity	6000 bottles/day (200 ml)
Land: 600 sq. m. with development charges	1.50 lakh
Building: Office 20 sq m.	0.40 lakh
Processing Area 80 sq. m	3.20 lakh
Storage space 40 sq m.	1.60 lakh
Auxillary constructions like platform, tube well, tanks etc	1.00 lakh
	7.70 lakh

Pre-cleaning of Sugarcane

↓

Crushing and filtering through filter cloth

↓

Addition of ingredients

↓

Flushing with CO_2

↓

Tapping clear juice

↓

Filtration

↓

Mixing preservatives and flavouring agents/juices

↓

Bottling

↓

Pasteurization

↓

Storage

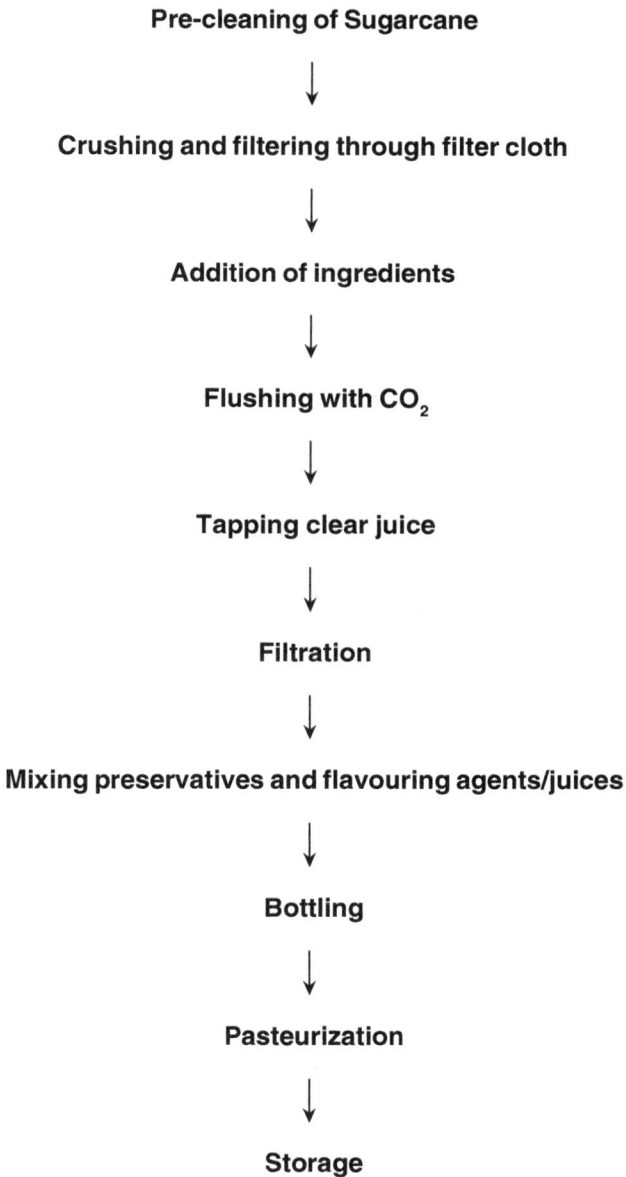

Process Flow Chart

Equipments Required

Main Equipments

☆ Sugarcane crusher

☆ S. S. Tanks (2)

☆ Bottle washing unit

☆ Bottle filling unit

☆ Plate and frame filter

☆ Crown corking machine

☆ Conveyor and Inspertion belt

☆ Pasteurizing tanks

☆ Walk in cooler/chamber at 10°C

☆ Carbonator

Auxiliary Equipments

☆ Water treatment plant

☆ Weighing scale

☆ Balance

☆ Pumps etc. ·

Other Assets

☆ Van-transport

☆ 60,000 bottles

☆ Gas cylinder,Crates

Raw Materials Required

Sugarcane, Preservatives, Flavouring agents, Carbon dioxide

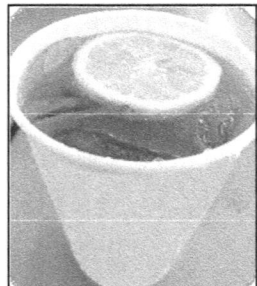

Utilities

Power 55K VA, Water- 7000 litres/day

Cost of Production

Equipment cost	15.00 lakh
Fixed cost	Rs 22.70 lakh
Working capital	7.00 lakh
Project cost	Rs. 29.70 (say 30.00 lakh)
Cost of production	30 lakhs
Sales (6000x225 days Rs. 3.00/bottle)	40 lakhs
Profit	40-30=10 lakhs
Return on investment	100x10/30.0= 33 per cent

Sugarcane juice is a nourishing 'Swadeshi Drink'. It should be promoted throughout the country, shelf life of bottle is three months. The investment could be scaled down by reducing the production capacity

Source of Technology

Technology is available from C.F.T.R.I., Mysore

19

Edible Soybean Flour

Soybean is consumed in China, Japan and other East Asian countries as a traditional food from time immemorial. It finds use in traditional foods in East Asian including Japan, Korea, Indonesia and Thailand as a richest and cheapest source of protein.

Soybean is an excellent raw material for a number of fabricated food products with desired composition, texture and flavour. In Uttarakhand, Himachal Pradesh and Assam

etc., soybean has been in use in a small way. Madhya Pradesh is the largest producer of soybean and a large number of solvent extraction plants are operating for extraction of soybean oil and de-oiled cake.

Constituents of Soybean

Fats	18-20 per cent
Protein	40-44 per cent
Carbohydrates	20-30 per cent
Minerals	(for 100 gm of raw material)
Ca	226 mg
P	546 mg
Fe	5 mg
Mg	276 mg
Cu	2-4 mg

Thus, if one consumes soybean daily there is no need to take any tonic.

Advantages of Soyflour

☆ Egestron harmone in soybean cures gyneocological problems in women.

☆ Controls blood pressure

☆ Lowers cholesterol

☆ Immunity to diabetes and arthritis

☆ Prevention against cancer

☆ Leads to slimness

Manufacturing Process

The process of manufacturing soyflour is very simple. First of all, the soybeans are cleaned and dehusked to remove

husk. It may be noted that husk should not be consumed. The dehusked seeds are blanched and water is allowed to drain off. The soybeans having (60-65 per cent moisture) are dried in sun or using hot air drier till moisture content of 6-7 per cent is achieved. The dried beans are ground in hand operated or power *chakkies*. Due to 18-20 per cent oil content in soybean the grinding capacity is drastically reduced as material sticks to burr plates which require frequent cleaning.

It is advised to grind mixture of wheat or other grains with treated soybean splits.

Soybean flour can be used for protein fortification of wheat flour. Addition of 500 gm of soyflour is recommended in 10 kg of wheat flour.

Typical Composition of Full Fat Soyflour Biscuits (Per cent)

Maida	100.00
Sugar	40.00
Vanaspati	1.0
Baking powder	40.00
Sodium-bicarbonate	0.80
Salt	1.00
Soy Component	30.00
Water	32.00

Soyflour is extensively used in meat products, cereals, ready-to-eat products, food, drinks, baby food, confectioneries, candy products, special diet food, high protein soups, protein concentrates, food additives etc.

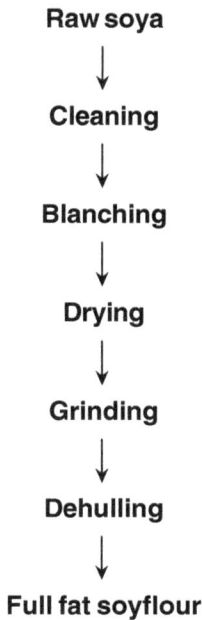

Raw soya

↓

Cleaning

↓

Blanching

↓

Drying

↓

Grinding

↓

Dehulling

↓

Full fat soyflour

Process Flow Chart

Equipment Required

☆ Cleaner

☆ Dehusker

☆ Blancher

☆ Drier

Capacity

	100 kg/day/shift
Building (on rental) 600 sq. ft	Rs. 2000
Equipment	Rs. 1.25 lakh
Working capital 25 days cost of raw material Rs. 120 kg/day @ Rs. 25/Kg	Rs. 75,000
Fuel/Electricity	Rs. 5,000
Man power (2 persons)	Rs. 8,000
Packaging material	Rs. 2000
Misc. Rs. 5000	
Cost of production	Rs. 95, 000
Sales Rs. 50/- per kg for 2500	Rs. 1.25 lakh
Profit (1,25,000- 95,000)	Rs. 30,000/per month.

☆ Grinder

☆ Weighing Scale

☆ Packaging Machine

20
Improved Jaggery (Gur) Making

Jaggery, commonly known as *'gur'* in India is a traditional, non-refined and centrifugal sugary material consumed in Asia, Africa, Latin America and Caribean countries.

It is rich in iron, a composite of haemoglobin which prevents anaemia (apart from nutrients and minerals). It is a source of energy for rural masses as well as it is consumed in urban areas during winter.

The main raw materials used in it are sugarcane and date palms. It is a traditional cottage industry in villages producing sugarcane and date palms.

Following steps are involved in the production of jaggery:

1. Cutting sugarcane from fields

2. Feeding in roller crushers

3. Transfer to open boiling pans *(kadai)* heated by underground *'Bhatti'* made from building bricks joined by clay mortor

4. Adding ingredients like sodium-bi-carbonate for clarification of juice

5. Transfer thick (hot) paste to trays or moulds

In conventional design the *'Bhattis'* are constructed using building bricks joined by clay mud and bricks are not used on the floor of kiln. Indian Institute of Petroleum (IIP) Dehradun, Uttarakhand had developed the improved design of *Bhatti* in which three open pans *(kadaies)* are kept over the kiln in sequence for boiling of sugarcane juice.

The dried baggase is fed from one end of the kiln while at the other end chimney is provided for release of smoke. At present, due to improper design of kiln and chimney, more fuel is required and dense smoke is released out of chimney due to incomplete combustion.

Cutting sugarcane

↓

Feeding to roller crusher

↓

Transfer to boiling pan

↓

Addition of bi-carbonate

↓

Transfer to moulds/trays

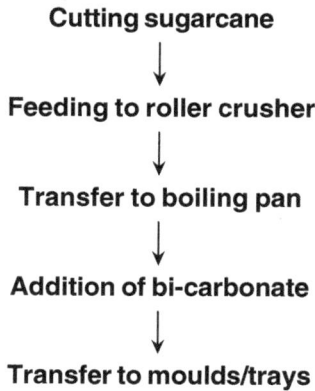

Process Flow Chart

Advantages of Improved Design

The improved design is eco-friendly and has the following advantages:

1. Saving of fuel (baggase) by 10 per cent

2. Reduction of smoke/gases from chimney as it is environmental friendly

3. Use of round chimney of appropriate height instead of square cross-section.

4. Use of refractory bricks enhances life of chimney

5. 20 per cent higher yield of gur from same quantity of juice

Sugarcane crusher

6. Improved quantity of *gur*
7. Improved fuel feeding system and air entry ports have been optimised for better combustion.
8. Bottom ash could be used as building material.

Techno-economics

1. Production capacity 8-10 q/day
2. Shed 400 sq. ft- 1.00 lakh

Equipment

1. Furnace and chimney
2. Power driven sugarcane crusher (electric motor or diesel engine)
3. Three boiling pans (MS 62 cm diameter of thickness 14,16, 22mm)
4. Moulds, trays etc.

Applications of *Gur*

Applications of *gur* have been diversified such as:

☆ Groundnut mixed *gur*

☆ Gur containing dry fruits cut into size

☆ Gur pieces containing ginger etc. for use in winter.

Gur Chocolate

The author has prepared chocolate toffee/bar using *gur* instead of sugar. It does not lead to dental cavities among children. The ingredients are *gur*, butter/pure ghee, milk powder and choco powder, etc.

Gur Powder

It is used in Punjab/western U.P. The *gur* in hot condition is powdered by rubbing with palms of hands (using gloves). It is yellowish in colour and is very tasty.

Packaging

Attractive packs could attract customers and products could be sold at higher selling prices.

21
Mini Rice Mill

India is the second largest producer of rice in the world. The mechanised sector of rice milling industry handles more than 45 million tonnes of paddy annually. About 10-15 million tonnes are processed through large scale modern mills involved in producing fine quality of rice for export market.

The major portion of paddy is still processed through hullers which are usually low capacity mills and result in very high percentage of brokens. In these hullers both shelling and polishing operations are carried out simultaneously and there is no control on the polishing of rice. As a result impure bran mixed with husk is obtained and a higher breakage of rice results in loss of revenue. To overcome these problems it is necessary to carry out the shelling and polishing in two separate units.

A mini rice mill has been deeloped by CFTRI, Mysore and many manufacturers have sprung up having their own versions. The sheller is a compact unit designed on the densimetric classification principle; the polisher could be light either vertical cone polisher or a horizantal roller polisher.

Salient features of mini rice mill

☆ 1-4 per cent extra yield of head rice depending on the variety

☆ Production of pure rice bran free from husk

☆ Control over degree of polishing from 3 per cent onwards

☆ Capacity as low as 500 kg of paddy/hour and can also serve as custom milling unit

The pure bran can be supplied to solvent extraction industry after stablisation for production of edible rice bran oil. Chemical stabilisation process is easy to operate compared to steam or roasting treatment. The principle of the process lies in maintaining pH of the bran to the level where the lipase activity would be negligble and within safe limit for Free Fatty Acids (FFA) which is 2-4 per cent for edible oil.

Technical Aspects

Location

There is a vast potential for installing mini rice mills in all paddy growing areas, as a rural small scale activity.

The mini rice mill consists of a paddy-cleaner, sheller, separator and a polisher. The separator is a compact unit

```
        ┌─────────────┐
        │    Paddy    │
        └─────────────┘

    ┌──────────────────────┐
    │  Cleaning- destoning │
    └──────────────────────┘
             │
             ▼
        ┌─────────────┐
        │   Shelling  │
        └─────────────┘
             │
             ▼
    ┌─────────────┐        ┌─────────────┐
    │ Separation  │ ────▶  │    Husk     │
    └─────────────┘        └─────────────┘
             │
             ▼
    ┌─────────────┐        ┌─────────────┐
    │  Polishing  │ ────▶  │    Bran     │
    └─────────────┘        └─────────────┘
             │                    │
             ▼                    ▼
    ┌─────────────┐        ┌──────────────┐
    │    Rice     │        │ Stabilization│
    └─────────────┘        └──────────────┘
                                  │
                                  ▼
                          ┌──────────────┐
                          │  Supply to   │
                          │ solvent plant│
                          └──────────────┘
```

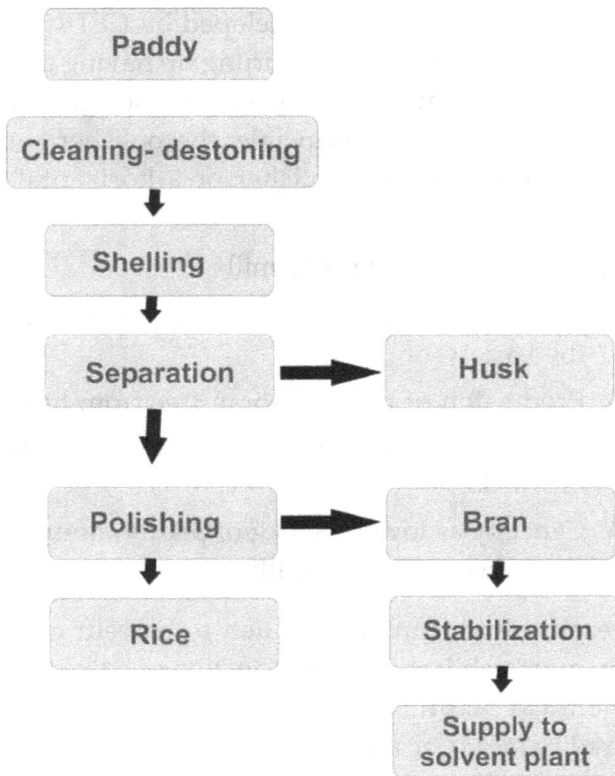

Process Flow Chart

designed on the densimetric classification principle. The polisher could be either a vertical cone polisher or a horizontal rotor polisher. Even a huller used for milling could serve as a polisher though there may be more breakage of rice.

The most important feature of the mill is that the shelling and polishing are kept separate. Because of the low capacity, a centrifugal sheller is most commonly employed.

Different types of units could be used as polisher. For maximum advantage, it is necessary to use a paddy separator, whereby need of a high polish can be avoided.

Quality Control and Standards: As per AGMARK specification

Suppliers of Machinery

☆ Mysore precision Engineers, C-123-124, Industrial Estate, Yadavgiri, Mysore

☆ Kisan Krishi Yantra Udyog, Kanpur, U.P.

☆ Vishwakarma Iron Foundry, Kashmiriya, Tanda, Dist. Ambedkar Nagar, U.P.

☆ Super Engineering Works, Howrah-Amta Road, Dass Nagar, Howrah

Economics

☆ Land 500 sqm.

☆ Work shed 100 sqm.

☆ Machinery - Rs. 3.00 lakh.

Tribals/villagers go to long distances to get their paddy processed for domestic consumption. They also resort manual processing where quality of rice produced is very poor. Entrepreneurs can set up mini rice mills as a service unit for custom milling. Mini rice mills are also viable in small paddy growing pockets of the country.

The Basmati rice could also be processed in mini rice mills with an attractive proposal.

In case of production of parboiled rice the paddy is soaked in water for 10-12 hours and then steam dried before transfer to sheller.

The machines are commercially available from manufacturers. Those interested in manufacturing of mini rice mills can contact CFTRI Mysore for design/Technology. To promote the improved rice milling, government is offering subsidies.

By Products

☆ Rice Bran for supply to solvent extraction plants after stabilisation

☆ Paddy husk- use in particle board industry, rice husk cement

Down line products like rice flakes 'poha' units could also be set up which offer good return.

Techno-economics

Production capacity	500 kg of paddy/hour or 4T/day
Land 500 sq m.	₹ 50,000
workshed + godown	₹ 2,00,000
Machinery	₹ 3,00,000
Pre-operations/Miscellaneous	₹ 50,000
Total	₹ 6,00,000

ENERGY

22
Biomass Gasifier

Biomass is basically an organic material which stores solar energy from sun through the process of photosynthesis. Biomass fuels include agricultural waste, crop residues, fuel wood, woody waste and organic waste like tapioca stalks, soyabean stalks, ground nut shell, coconut shells, sugarcane trash and corn stalks/cobs.

Biomass gasifiers are eco-friendly and renewable source of energy. Biomass does not release CO_2 to the atmosphere unlike fossil fuels as it absorbs the same amount of carbon during growth.

Biomass gasifier converts solid biomass into more convenient gaseous form. Normally, the ratio of air to fuel required for complete combustion of the biomass is 6:1 to 6.5:1 which is defined as stoichiometric combustion with end products being CO_2 and H_2O. In the gasifier the air-

fuel ratio is 1.5:1 to 1.8:1 under sub stoichiometric conditions. The gas produced is producer gas which is combustible having calorific value of 4.5-5.0 MJ/kg.

The Gasifier System

The gasifier system comprises of a reactor for generation of combustible producer gas which after cleaning and

cooling process is made available for power generation/ thermal applications. The producer gas from gassifier can be used to run diesel engine. Gasifier when connected with electric generator through thermal engine can generate electricity. One cu.m. of gas can generate thermal energy of the order of 4.5-5.0 Mj with flame temperature of upto 1200°C.

The Gasification Process

The gasification process converts carboneceous material such as coal, petrol and bio-fuel, biomass into carbon mono-oxide (CO) and hydrogen (H). Gasification is more effective than direct combustion of the original fuel because it can be combusted at higher temperature so that the thermo-dynamic upper limit to the efficiency defined by 'carnot cycle' is higher. Higher temperature combustion refines out corrosive elements such as chloride and potassium. During pyrolysis process carboneceous matter heats up and volatile matter are released and char is produced resulting in 70 per cent weight loss. When gasification process occurs char reacts with steam producing CO and H_2.

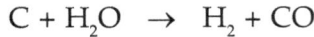

$$C + H_2O \rightarrow H_2 + CO$$

The Gasifier Process

In the 19th century gasification was used for production of synthetic chemicals. Sustainable plants of capacity 250-1000 KWhr and have been set up for rural electrification in remote villages which were not connected by State Electricity Company. There are no tansmission line losses and desired voltage is obtained due to localised supply.

About 1.2 kg of wood/biomass is required for generation of 1.0 kwhr of energy. About 20-30 kwh plants which are sufficient for needs of villages cost around Rs. 2.00-3.00 lakhs, including cost of electricity generation. One plant has been set-up at Kukru Khamla near Betul (M.P.) where a *'Atta chakki'* is being run through gasifier. In case of higher capacity plants like 2.5-6.00 mega watt there is provision of supply to State Electricity Grid after signing MoU with State Electricity Board.

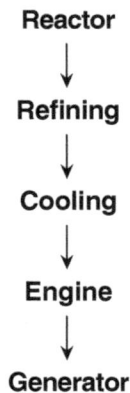

Reactor

↓

Refining

↓

Cooling

↓

Engine

↓

Generator

Process Flow Chart

Finance and subsidy etc. is provided by Ministry of Non-conventional and Renewable Energy Sources, Government of India, New Delhi.

Entrepreneurs can contact State Energy Corporation of their State for assistance.

23

Low Cost Briquetted Fuel

Briquetted fuel is the answer to the ever-increasing energy crisis. This unique alternative to natural coal is easily manufactured from agricultural and forest wastes. With a calorific value of about 4000 kcal/kg, the raw materials give superior briquettes particularly because of negligible ash content while burning. The manufacturing process is extremely simple. Raw materials fed into a hopper are forced-fed between the punch and die, resulting in high compression and hence high temperature. Cylindrical briquettes are formed owing to carbonization achieved by hardening of surface. Saw dust, coir waste (pith), groundnut shells, rice husk (de-oiled rice bran), pine

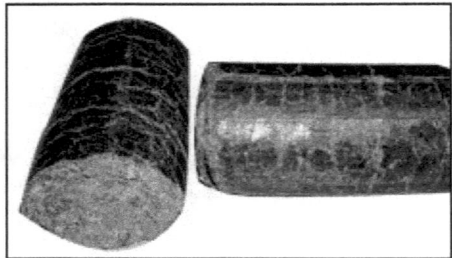

needles (pulverised), hardwood shaving, softwood shaving, bagasse (pith), bark (pulverised) and plant leaves, etc. can be briquetted.

In India at least 150 million tons of forestry wastes and 350 million tons of agriculture residues are available every year, which are not being utilized. A substantial portion of these residues goes waste and this even causes environmental pollution. At the outskirts of every Indian village heaps of agricultural wastes generate air pollution. On the other hand there is scarcity of fuel for domestic and commercial needs in the rural areas, which result in indiscriminate felling of trees in the forest areas. People, particularly women carry head loads of wood several kilometers to burn their *chullas*. It is possible to prevent or at least reduce the deforestation due to fuel needs by providing alternative fuel as traditional fuel like coal,

kerosene and gas etc. are beyond the reach of the people in rural or forest areas. Utilization of agro/forest wastes by conversion into fuel would help in resolving fuel problems.

Several technologies have been developed for conversion of biomass into fuel such as compaction of agro-waste into briquettes taking advantage of lignin present in the biomass as binder. But this technology has not become popular due to use of high pressure extruders/hydraulic presses, which require high-energy input to operate these machines. Fluidized bed combustion/charring of biomass and briquetting the charred biomass using external binder are feasible.

Numerous charcoal production systems and improved techniques have been developed and are available for application from cottage level to commercial level using drum and mould methods to brick beehives, retorts and extruders. Appropriate process route has to be identified keeping in view the socio-economic and ecological factors.

Raw Materials

Some items that can be briquetted include bamboo chips, peanut husk, coffee bean husk and other vegetable wastes. Spent fruits from juice making operations such as cashew apples, mango seeds and hulls, etc may also be dried and briquetted. Any type of Agro-Forestry waste can be used. But the moisture content should be less than 12 per cent, grain size is below 5×5mm. Groundnut-shell, sugarcane biogases, caster shells/stalk, saw dust, coffee husk, paddy straw, sunflower stalk, cotton stalks, tobacco waste mustard stalk, jute waste, bamboo dust, tea waste, wheat straw, palm husk, soybean husk, coir pitch bark, straws, rice husk, forestry wastes, wood chips and many other agro wastes, can be used for the purpose.

An Excellent Enterprise

Developing a small business of producing fuel briquettes is an excellent way to develop a successful business. The enterprise can be begun with simple equipment and developed into a large scale business in just a few years. The product can be initially sold within the country before resorting to full fledged export. It can be first sold in bulk and later expanded to shrink wrapping and other consumer packaging of the products.

The enterprise is lucrative for several reasons:

☆ Raw materials are readily available in the country

☆ Low cost machinery can be used

☆ One may start small and grow rapidly

☆ Raw materials can be grown individually and also purchased cheaply in the country

☆ Minimum scientific knowhow is required to manufacture fuel briquettes

☆ Minimum industrial infrastructure is required to manufacture fuel briquettes

☆ The fuel briquette manufacturing project can be labour intensive and employ many people in the local area

☆ The risks as an entrepreneur are minimum than if one goes into a large scale project to begin

☆ Local repair and maintenance facilities can be used

☆ A fuel briquetting plant can be relatively inexpensive to purchase and operate

☆ A fuel briquetting plant can use lower skill level labour inputs

☆ A fuel briquetting plant requires lower technical skill levels for one as an entrepreneur

☆ The project can be upgraded with the development of the infrastructure in the country

☆ The enterprise can be the basis for rural agro-industrial development

☆ A fuel briquetting plant can inspire local craftsmen to innovate and develop improvements in the technology

☆ Fuel briquette manufacturing can produce products for country use as well as for export and urban use.

☆ The enterprise can reduce the use of expensive imported fuel.

☆ Fuel briquetting can help fight the greenhouse effect by avoiding the excess use of fossil fuel.

☆ A fuel briquetting plant can conserve valuable resources by utilizing wastes.

☆ Export markets for fuel briquettes have unlimited potential for profit

Each briquette is of 175/180 mm dia. Weight of one briquette is about 300 gm which can be ignited in a ordinary clay *chullah* of round shape having two steel bars for supporting the briquettes. It lasts for 40-60 minutes for cooking purpose and about 1.5 hr for heating purpose and gives sustained and smokeless combustion. Each of the 19 holes in briquette act as a tiny gasifier and briquette once ignited provides nineteen distributed flames just like LPG

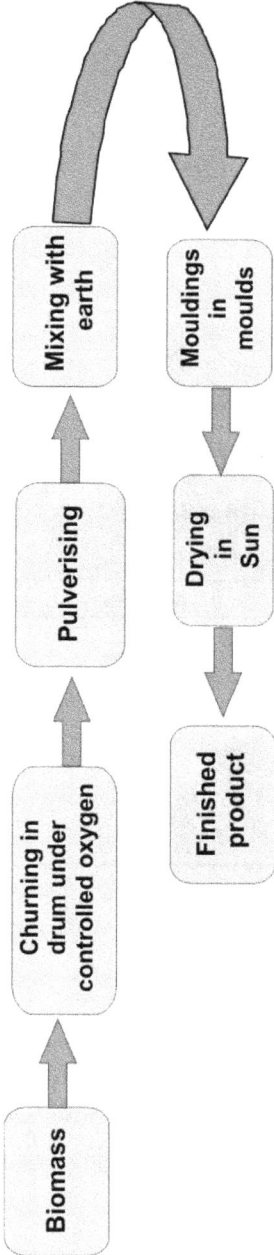

Flowchart of the Charcoal Briquetting Process

stove. Two briquettes (one over other) can also be used if the *chullha* is made to accommodate them.

Calorific value of briquette is 7000-7500 kcal/kg

Moisture	7-8 per cent
Volatile matter	13-14 per cent
Fixed Carbon	80 per cent

Capacity

22,500 briquettes/year of standard size or 150 briquettes per day for 150 days per year.

Machinery Equipment

Following minimum equipment is required for a cottage scale unit:

Equipment	Number	Cost (Rs.)
Drums	2	6,000
Stove	1	2,500
Moulds	2	2,000
Mixer/Misc. items	1	4,000
Tools/Misc. items		6,000
		Rs. 20,500.00

Power Requirement

It is a hand operated unit and no electric power is required to operate the unit.

Raw Material

Biomass 150 kg/day for 150 days

Or

22,500 kg/year

Local clay-3000 kg/year

(20 kg per day for 150 days)

Others

Small quantities of water are required for moisturizing the mix.

Land	800 Sq.ft.
Low cost shed 250 Sq.ft.	Rs. 30,000
Machinery	Rs. 20,000
Working capital	Rs. 60,000
Profit	Rs. 60,000 per year

In South India, instead of clay, cassava or starch (5 to 6 per cent) is used as binder and press could be used for making charcoal briquettes. This charcoal has several advantages over hard wood charcoals and used for Industrial applications such as manufacture of carbon disulphide, carbon electrodes, carbon tetra chloride, sodium cyanide, activated charcoal, industrial ovens, water purifiers, and electrodes for primary batteries.

For further information entrepreneurs can contact following Research Institute:

☆ AMM Murragappa Chettier Research Centre, Taramani, Chennai.

☆ M/s Development Alternatives, Orchha, Jhansi.

24
Solar Devices

Introduction

The need for alternative sources of energy has become of prime importance due to depletion of fossil fuels. Among alternative sources available in the world the solar energy offers immense potential as in most parts of country sun shine is available for more than 275 days in a year. Solar energy is a pollution free source even comparable to nuclear power plants due to problems in disposal of spent fuel. Considerable Research and Development work has been carried out for utilization of "Solar Energy" in Research laboratories, PSUs, Universities etc. but still it has not at taken desirable pace. Devices like solar water heaters, solar cookers and ovens, solar space heating systems, solar driers, solar desalination plants, power plants, solar refrigeration systems etc. have been developed. There is dire need to introduce solar devices in day to day life. Utility products like solar water heaters should be included in the design/plan of houses/buildings.

Thermal Devices

The heart of a solar system is solar "collector". The efficiency of a system depends upon the type of collector used. Several types of collectors have been developed by different organizations, such as:

1. Copper tubes brazed with aluminum sheet.
2. Copper tubes brazed with copper sheet. No doubt it is costly but gives best thermal efficiency.
3. Aluminum tubes fixed to aluminum sheet.
4. Two corrugated galvanized sheets bonded together to form tubes.
5. Bond ducting using pressed aluminum sheets.

The panels are coated with black selective coating to achieve very high absorption efficiency. During early days dull black paint was used to coat the panels. Subsequently, rare metal oxides were added to the paint CECRI. Karaikudi and NAL Bangalore have developed black nickel electrolytic coating which enhances efficiency by 40 per cent. The flat plate collector consists of sheet and tube panel coated with selective black coating kept in an insulated box. The box is fitted with glass sheet on the top side and insulated at the bottom. Water enters the collector through a header tube and passes through the riser tube and comes out on the top through another header tube. The water rises due to thermo-siphoning effect but to accelerate flow rate, fractional HP pumps are also provided. The collector is facing south so that sun shine is received throughout the day (E to W). Collector of 0.5x2m (1.0 sq.meter) area can heat 100 L water to 60-85° C depending on the specification of material and selective coating used. These flat plate collectors with pipe line and insulated tank are available at the cost of Rs. 20000/- per meter sq.area. Author was associated with 10000 litre per day solar water heating system at Astha and Bhopal in M.P., Dugdh Maha Sangh during eightees. CSMCRI Bhavnagar installed a water heating system at Ahmedabad to pre heat boiler feed water in a textile mill in eightees. In late seventees author was involved in the installation of solar water heating system for a swimming pool in a five star hotel in Delhi. These examples show enormous potential of solar energy to save fuel. It may be added that for use during cloudy/rainy days, electric heating elements are provided in the water tank. In Chinese design concentric glass tubes are used. Inner tube is coated with black selective coating and gap between the tubes is having vacuum to avoid radiation of heat.

Solar Cookers

Solar cookers fabricated from steel sheets in parabolic shape concentrate heat energy at the focus of parabola/ paraboliod. These were developed by National Physical Laboratory, New Delhi in fifties. It enabled making chapattis, frying, cooking etc. but house wife has to be in sun. Subsequently, to increase efficiency glass was moulded in an oven in the shape of paraboloid. The back side of the glass was silvered and provided protective coating to retain silvered surface. It enabled to get 1 kw heat energy from 1.0 meter diameter paraboloid at its focus. A NGO at Indore has designed/fabricated and setup a prabolic cookers with focus inside the kitchen. They used steel fins imported from West Germany, Heliostats have also been fabricated by them to change position of cooker with the position of sun. They cook meal, for 40 persons daily (dia 3.0 metre).

The use of box type solar cooker is well-known. Modifications are required to improve efficiency and to reduce cooking time. By incorporating selective coating on black box and anti-reflection coating on transparent glass, we can increase efficiency by 40 per cent.

Solar Oven

Solar ovens with plane booster mirrors have been fabricated. This oven is provided with an octagonal arrangement of booster mirrors to enhance concentration ratio and temperature upto 250°C. This oven can be used as domestic bakery. A cake is produced is 40-45 minutes.

Solar Driers

Various organizations have developed solar driers for drying of fruits and vegetables. It offers drying of food stuffs in clean atmosphere at a temperature of 20-25°C above ambient. Thus drying time in reduced considerably. Commercial models could be fabricated using locally available materials.

Solar Stills

Drinking water is a big problem in coastal and high salinity areas, women have to fetch water from long distances. CSMCRI Bhavnagar and AMM. Murugappa Chetiar Research Institute, Chennai have developed solar stills for purification of drinking water to make it potable. It is possible to obtain 100L/day water by installing solar stills costing Rs. 7000/-, Author can provide the design. Solar stills are also suitable for making battery water 100L per day worth about Rs. 1000 per day earning for entrepreneur. For solar power stations, cylindrical paraboloids made of stainless steel are used as collectors. On the focus of these collectors tubing is provide where steam is produced and then compressed and used to run turbine. MNES has initiated several projects for setting up solar power stations. The power could be fed to a grid or small solar power houses could be setup to meet need of local population.

Solar Photovoltaic Cell (SPV) System

In a photovoltaic cell photons in sunlight hit solar panel and absorbed by semi-conducting materials like silicon. Electrons are knocked down loose from atoms, allowing them to flow the material generating electricity. The semi conducting materials used are:

1. Single crystal and polycrystalline silicon.
2. Amorphous silicon/silicon thin films.
3. Cadmium telluride and sullphide
4. Copper indium selenide.
5. Galium arsenide.
6. Light absorbing dyes.
7. Organo-polymer cells.

Various Universities, IITs and Research Laboratories like NPL New Delhi, NCL Pune, BARC Mumbai etc. carried out R and D to develop solar cells and solar cells are being manufactured by M/s. Central Electronics Limited (CEL) Gaziabad. Entrepreneurs can buy solar cells and assemble panels which are kept under glass cover/frame to prevent from dust, rain etc. These need to be cleaned periodically.

Solar cells are modular in nature and these could be connected in series/parallel to enhance power capacity and these are eco-friendly as they do not consume fossil fuels and are easy to install, operate and maintain. The applications include solar lanterns, solar street light, domestic lights, water pumping systems and solar power generation.

A compact lighting system consists of CFL lamp (7-8W), a battery power pack and an electronic circuitry housed in a metallic/plastic casing and a PV panel. The lantern operates for 3-4 hours once charged in a day for 5 hours in sun. It is a boon for remote villages where regular electricity is not available. It can work as emergency light in cities. A typical system consists of 37 W solar PV model, a 12 V, 40 Ah low maintenance tubular battery and control system. The inverter operates for 4 hours when charged during the day. Solar inverters could he marketed to replace conventional models used.

There is a need to popularize solar devices to supplement our power supply system. Ministry of New and Renewal Energy, Government of India offers attractive subsidies to promote solar and other alternative energy systems.

Entrepreneurs could set up a sheet metal fabrication workshop with black nickel coating plant for manufacturing "Solar Panels" technology for Black Nickel coating could be obtained form CECRI, Karaikudi, Tamil Nadu.

25
Purification of Water by Solar Energy

The current boom in population in urban areas and the increasing intensive irrigation requirements for agriculture in rural regions have already resulted in considerable drying up of wells and aquifers.

The decreasing availability of water has necessitated in the search for fresh sources of drinking water. The available water in many areas in the country is brackish, saline or impure. Salinity is a major problem in the coastal areas.

Though efforts are made to supply water through public distribution systems there are inherent limitations to this programme.

Processes Available for Purification of Water

In our country pure drinking water is a major problem in tribal/rural areas. There are many processes available for purification of drinking water. These processes are:

☆ Chlorine tablets

☆ Pot chlorination of wells

☆ Slow and rapid sand filters

☆ Water filter candles

☆ Removal of iron by directly connecting to hand pump

☆ Fluoride removal

☆ Reverse osmosis plants/electrical or bullock driven plants

☆ Solar stills

☆ Arsenic removal plants, etc.

Purification of Water by Solar Stills

Central Salt and Marine Chemical Research Institute (CSMCRI), Bhavnagar, Gujarat, purified water by the use of solar energy which is abundantly available from 275-300 days in a year and has installed solar stills in various parts of Kutch and Gujarat.

AMM Muragappa Chettiar Research Centre of Chennai has also developed designs of solar stills to suit the rural economy. Application of solar energy are well known like solar water heaters, space/room heating systems, solar ovens, heating of boiler feed water in industry, box and parabolic cookers, solar cells for lighting and solar pumps, etc.

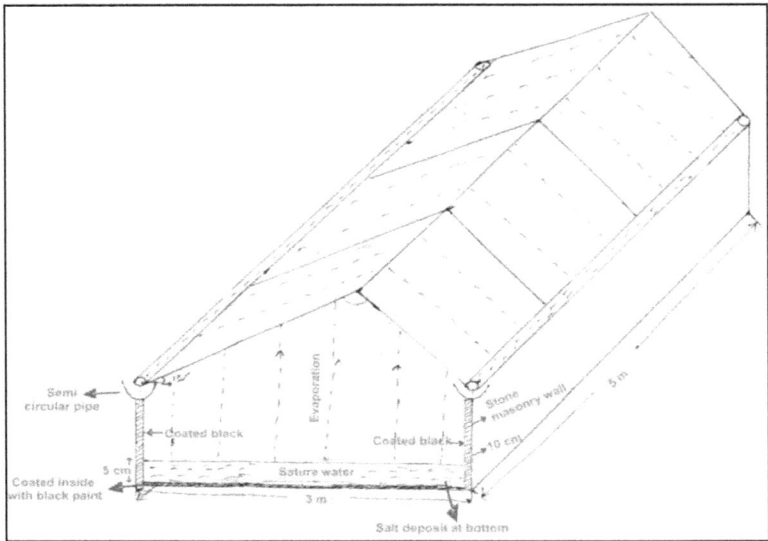

Solar still

Solar still is very useful for purification of water by desalination in the stills and exposure to UV rays of sun.

This is easily applicable in rural and urban India alike as the energy requirement for the distillation units is met from solar radiation. India is gifted with rich sunshine and therefore solar energy is not limit here.

MCRC's involvement in fabrication of solar stills started in 1979 with the basic conical type named "Thoyam".

In Thoyam a pan of impure water is enclosed in a transparent cone-shaped cover and traps solar heat causing part of the heated water to evaporate, rise to the top and condense on the cooler transparent surface inside the cover.

The pure water thus formed drips through a channel to a reservoir. This set up fabricated from inexpensive and locally available materials, can produce 3 to 4 litres/m²/

day in bright sun light. Field trials in the MCRC has designed new solar stills through its RandD processes to yield 8 to 10 litres of potable water/m²/day, which is sufficient for a family.

Water produced by solar stills could also be used as battery water. Entrepreneurs can produce battery water and supply it to the market.

These could be felicitated by artisans available in rural areas. Layout of the still is given below.

Materials Required

☆ Window glass

☆ Stones/bricks

☆ Dull black paint

☆ Angle iron frames

☆ Semi-circular pipe

☆ GI Sheet, etc.

Investment

Almost no cost is involved. Only the cost for installation is required, which is - Rs. 5,000 to Rs. 6,000 for producing 100 lit. of water.

The Selling Price

The selling price of 100 lit. of water could be Rs. 100x10 = Rs 1,000.

26
Bio-diesel from Jatropha

India has a landmass of 325 million hectares out of which 65 million is wasteland. These areas are scattered and fragile and suffer from prolonged droughts. Due to de-forestation soil becomes susceptible to erosion resulting in decreasing top soil and lowering of ground water level. These wastelands could be economically utilised by cultivating species like Jatropha curcus or Salvadora for bio-diesel production as an economic activity. The Jatropha Curcus (L) or physic nut is a drought resistant and multipurpose plant not grazed by animals. Although a native of tropical (South) America, it thrives well in Africa and Asia (India as well). It grows well in tropical and sub-tropical regions of the world having low rainfall and problematic soils. Being drought resistant it can help to reclaim barren areas and could be grown as boundry fence/ hedge to augment income.

Tribals of Baster have been using seeds of Jatropha as a torch by buring seeds. Due to their oil content, seeds are made to pass through a wire which continues to burn in the night. They are also using Jatropha oil in engines (prime movers) by the adjusting the ejector nozzle for higher viscocity oil of Jatropha. They contain 32-33 per cent (by weight) viscous oil which can be used to manufacture candles and soap in cosmetic industry, for cooking or as diesel paraffin substitute or extender in the paint industry.

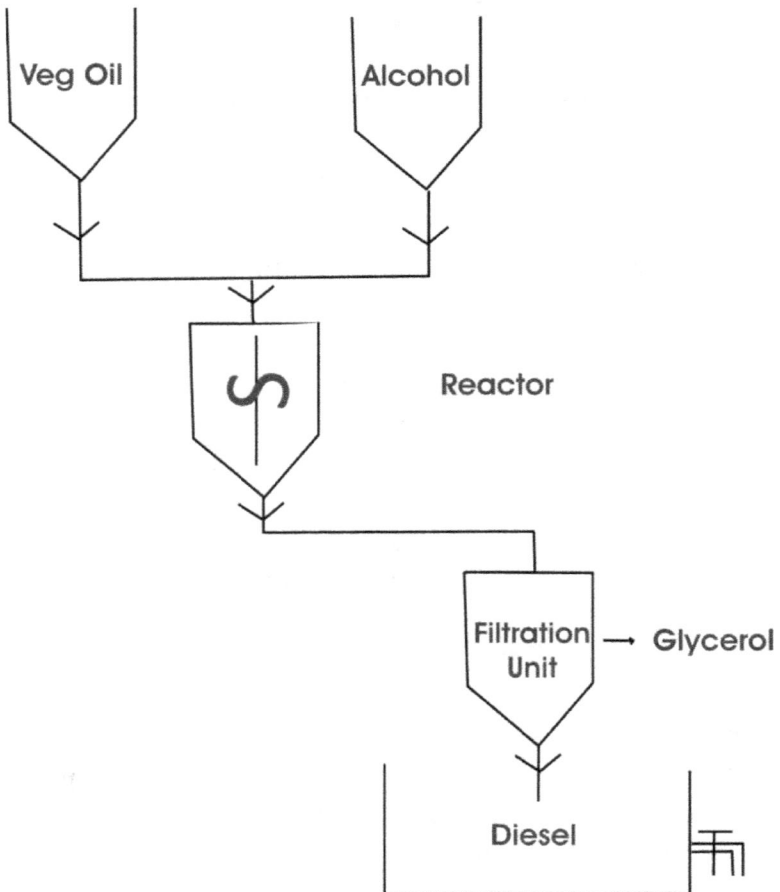

Veg Oil

Alcohol

Reactor

Filtration Unit → Glycerol

Diesel

**Raw materials –Veg oil + methanol
NaOH/KOH as catalyst**

↓

**Reaction in a reactor heating (60-65°C) and involving
trans-estrification 2-2- ½ hrs**

↓

Filtration ⟶ **Glycerol**

↓

Biodiesel

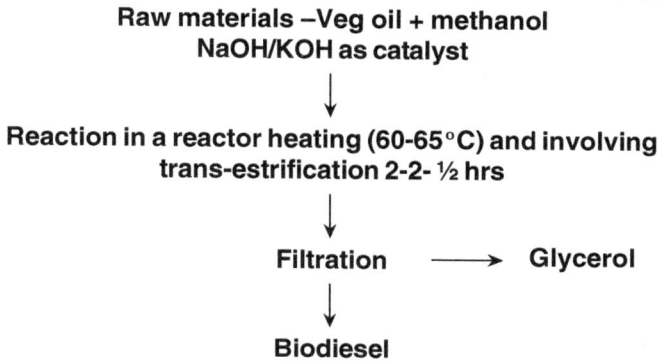

Process flow chart

Equipments

1. Reactor with heating and revolving arrangement
2. Filteration unit
3. Pumps

Biodiesel is automotive fuel of future and it can best supplement the existing sources. The main oil seeds used for production of diesel includes *Jatropha curcus* (*Ratan Jot*), *Madhuca latifora* (*Mahua*), *Salva dora pezsica* (*Pilu*) and *Pongamia pinnate* (*Karauj*). These plant species are found in abundance in Gujarat, Andhra Pradesh and Madhya Pradesh. Some farmers and entrepreneurs have started cultivation of *Jatropha curcus* for employment generation and biofuel.

Process

Biodiesel is produced by chemically reacting vegetable oil/animal fat with alcohol to produce a new compound called fatty acid alkyl ester by trans-estrification. A catalyst such as sodium/potassium hydroxide is used and glycerol is obtained as by-product.

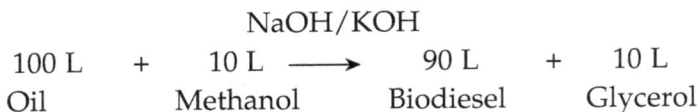

$$\begin{array}{ccccccc}
& & \text{NaOH/KOH} & & & & \\
100\text{ L} & + & 10\text{ L} & \longrightarrow & 90\text{ L} & + & 10\text{ L} \\
\text{Oil} & & \text{Methanol} & & \text{Biodiesel} & & \text{Glycerol}
\end{array}$$

Methanol is recovered in the process.

Agro-technology

Soil: Wasteland unattractive for agriculture due to low fertility and alkaline soils are used.

Rainfall: 200- 1200mm in hot climatic conditions.

Planting: Seeds or rooted cuttings are planted at a distance of 2m x 1.5m. About 2000 plants/ha or 400-500 plants/Km are planted

Yield: 0.4T/ha/yr after 2-3 years and 2.5- 3T/ha/yr after 5 years. Plant life is about 25 years.

Return	3.00 T x Rs. 6/kg = Rs. 18000/ha/yr or 25 years
Project cost	Rs. 6.00 lakhs
Capacity	400 L/day
Production cost	Rs. 30/- = Litre (one litre diesel from 3 kg of seed).

WASTE UTILIZATION

27
Utilisation of Fly Ash

Coal continues to play a major role in power generation although Government is giving stress to nuclear power and hydro-generation is contributing its share. At present 290 million tonne of coal is consumed every year which constitutes nearly 40 per cent of total power generation

which in turn produces 112 million tonne of fly ash and this is expected to increase to 175 million tonne by 2012 and 225 MT by 2017. This large volume of fly ash posseses serious environmental problems. As such, there is need to develop technologies for production of value added products on sustainable basis. Realising the magnanimity of problem GOI has set up Fly Ash Mission under Department of Science and Technology to tackle the problem in mission mode.

Some of the CSIR Laboratories like Central Building Research Institute (CBRI) Roorkee, Central Fuel Research Institute (CFRI) Dhanbad, Indian Institute of Minerals and Materials, Bhubaneshwar, RRL (AMPRI) Bhopal, Central Glass and Ceramic Research Institute (CGCRI) Kolkata and Central Power Research Institute (CPRI) Bangalore intiated Research and Development Projects more than two decades back and developed a couple of technologies.

First plant for production of 3,0000 bricks/day was set up near Kolkata by CFRI Dhanbad. There were manufacturers of machines for manufacturing of refractory bricks who switched over to manufacturing of machinery for fly ash bricks. CFRI set up second plant at Bandel (near Kolkata) with German machinery. Simultaneously CBRI, Roorkee also came out with technology for manufacturing of fly ash bricks and set up a plant at Okhla (New Delhi) to utilise fly ash generated by the nearby plant. They also developed clay fly ash bricks by partially replacing clay with fly ash which improved compresssive strength of bricks and also helped in fuel consumption of fly ash bricks produced by these plants. These are having compressive strength of 100-120 kg/cm^2 AMPRI/RRL, Bhopal also developed clay-fly ash bricks (red bricks) and transferred the technology, to local brick kilns who produced these bricks and demo-houses were constructed using clay fly ash bricks.

RRL/IIMM (Bhubaneshwar) developed the technology for production of fly ash aggregate by nodulising fly ash and sintering it at high temperature. These aggregates could be used for construction of roads in areas around Thermal Power Stations. CGCRI, Kolkata have also developed the technology for production of fly ash tiles.

Significant contribution has been made by a private firm of Vishakhapatnam who have at their own developed 'FALGI' process for production of bricks and they have set up 70-80 plants in the country having production capacity of 10,000 bricks/day and investment of 10-15 lakhs, it has been widely adopted.

Excessive soil degradation and reducing soil fertility is a continuing problem in many agricultural areas of the world and increasing dosage of fertilisers is a common phenomenon. As such cost of agricultural inputs has increased which can be reduced by use of fly ash which is a residue of burning coal and lignite. The micro and macro nutrients present in the coal generally get concentrated in the ash. By virue of nutrients it has the ability to modify the physical properties of soils. It works as a soil conditioner enhancing the yield of crops.

Under Fly Ash Mission demonstration projects have been undertaken at 50 locations having varied agro-climatic conditions. The large scale use of fly ash in agriculture and wasteland development has a potential to increase the yield by 15 per cent of grains, oil seeds, cotton and about 25-30 per cent of vegetables resulting in another green revolution. For *e.g.*, in case of marigold yield of flowers has increased by 10 per cent. This could be tried for other aromatic and medicinal plants as well as agro-forestry.

Use of Fly Ash in hydro-power and water resources sector is now being promoted. In the construction of dams for hydro power and water resources cost has been reduced substantialy. In cement production use of fly ash upto 18 per cent is permissible as per IS 456-2000.

Fly ash based Geopolymer concrete is an innovation material for the construction as it is free from hydration process, alkali and aggregate reaction and found highly durable in aggressive environment. Fly ash is rich in Silica (Si) and Aluminium (Al). It is activated using highly alkaline solution which will replace conventional cement in concrete mortar. Roller compacted concrete (RCC) is the key material of modern dam construction and pavement structures.

Fly ash is being used for construction of rural roads and their embankments. Central Road Research Institute has promoted use of fly ash in Road Construction at different locations.

Also ash has been used in construction of buildings in preperation of mortar for joinery work. A novel process has been developed by AMPRI, Bhopal for road construction using fly ash. In the preperation of pesticide dust for spray in field soft stone powders are used. It is suggested to substitute stone dust with fly ash. And lastly, artificial hills using fly ash can also be created on which greenery could be grown. As such fly ash can be used for landscaping.

Fly Ash Bricks

CBRI Roorkee has developed a small brick moulding machine to suit micro-level entrepreneurship and it is available to entrepreneurs through CBRI Roorkee.

Fly ash (70 per cent), sand (20 per cent) and lime (10 per cent) is mixed in a mixer with addition of water. The mix is transferred to vibro-press where bricks are moulded in a single operation. Subsequently the cast bricks are dried in sun and cured in water tanks for 21 days.

Techno-Economics

☆ Capacity- 3000 bricks/day

☆ Investment- Rs. 3.00 lakh

☆ Cost of machine- Rs. 80,000

☆ Mixer- Rs. 25,000

☆ Labour

Skilled- 1

Unskilled- 10

☆ Land- 700 sq. meter

☆ Power connector- 3KW

☆ Size- 100x100x200 mm.

☆ Colour- grey/cement

☆ Density- 1.6-1.7 gm/cu.cm.

☆ Crushing strength- 80-100 kg/cm^2

☆ Water absorption- 15-20 per cent

☆ Cost of production- Rs. 2500/thousand

☆ Selling price- Rs. 3500/thousand

The fly ash bricks save agricultural soil used for making conventional bricks. Some states have made use of fly-ash mandatory for producing fly ash clay bricks. Fly ash bricks do not require external plasting.

Use of fly ash would also help in saving the environment from the most dangerous pollution caused by Thermal Power Plants. Units should be set up preferably within 100 km radius of Power Plants.

28
Hand Made Paper

Conventionally paper is made by utilizing wood. However, excessive usage of wood for pulp and paper making results in denudation of forests leading to environmental problems. To avoid such a situation cellulosic sources (non-woody substrata) have to be resorted. These alternative substrata can be agricultural wastes such as sugarcane bagasse or straw belonging to paddy, wheat, maize, barley, ragi, millets etc. silk cotton (from *Ceiba pentandra*) has also been used for pulp and paper making. Here, the plant is undisturbed but only the silk cotton is harvested from the fallen pods, which is renewable. In addition to these alternate sources, making of handmade paper recycles the waste paper available in plenty.

Eco-Friendly Paper Making

Conventional method of paper making uses cellulose in the form of wood chips obtained by felling trees, leading

to increased global warming. Moreover, in this method, chemicals are used which release toxic substances causing environmental pollution (air, water and noise pollution) requiring treatment of water. Surveys state that for manufacturing 1 ton of paper 277 *Eucalyptus* or 462 bamboo plants are required. Moreover, 1,00,000 greeting cards made with handmade paper can save 500 trees.

Raw Material Selection

Used white paper is the best material; glossy papers are not useful for eco-friendly handmade paper making. Glue and staple pins are to be removed from paper before soaking. Tear the paper and soak the pieces overnight in water.

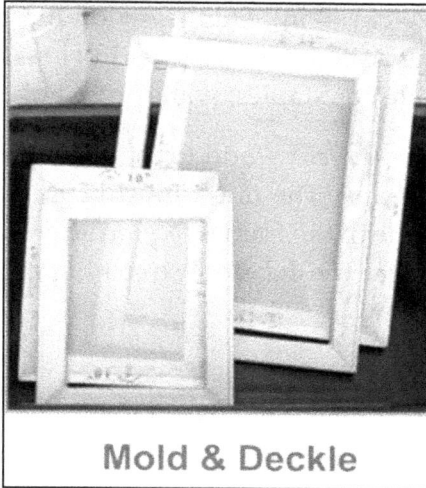

Mold & Deckle

Process of Production of Handmade Paper

Mold and Deckle

This simple device could be made easily with a carpenter's help. The mold comprises of a rectangular wooden frame. A layer of mesh (nylon net) is stretched on this mould.

The deckle (top portion) made of wood, is of the same size as the mould and is an open frame that rests on top of the mold. The mold is to be attached firmly with deckle by hook and eye clips or press-stud. The wet pulp is deposited onto the mesh before it is pressed and dried into sheets of paper.

Prepare the mold and deckle as follows:

Take 180 cm (6ft) length of 2cm (3/4 in) square wood and cut into,

1. Four 25 cm (10 in) lengths and

2. Four 20 cm (8 in) lengths; nylon mesh with 12 to 20 holes per cm (50 holes per in); brass or stainless steel pins, or staples; waterproof adhesive; nails.

Now arrange the cut wood to make two frames of the same size and shape. The mesh must be stretched tightly over the top of one of the frames. Nylon stretches when it is wet, so dampen the mesh before stretching it. Excess mesh should be trimmed.

Pulping

The waste raw materials soaked in the water is blended in a mixer to get the pulp.

Formation of Paper

The cuddaph or cement tank with one third-water should be kept ready for placing the mould and deckle unit. The pulp should be stirred well before placing on the VAT. The mould portion should be submerged in water when pulp is poured. The pulp is to be formed evenly on the VAT; the VAT is firmly held while lifting in order to allow the water to drain back through mesh. The thickness of the sheet is determined by the quantity of pulp.

Separation of Sheets

Separate the mould and deckle by unlocking the hooks. The evenly formed wet sheet/pulp on the mould is removed/separated by placing *gada* cloth uniformly over the pulp. This process is repeated to make many sheets.

Screw Press

The evenly formed wet sheets on the hand vat are removed by placing gada cloth interfacing over the finished sheet and this process is repeated to make subsequent sheets. This wet paper contains moisture and air bubbles. The water remain in the wet paper has to be removed thoroughly by applying even pressure on the paper so as to obtain smooth surfaced paper. The cloths with wet paper are stacked in the screw press.

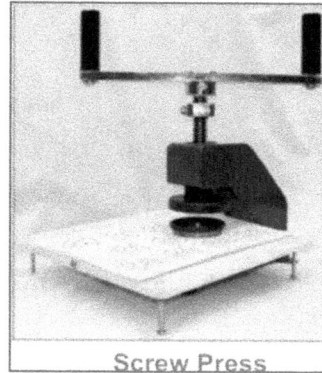

Screw Press

Drying and Calendering

The evenly formed wet sheets on gada cloth are allowed to dry by hanging on the ropes with clips. The sheets can be removed from the gada cloth once they are dried. The dried sheets are calendered, with the electric iron. The calendered sheets are then trimmed.

Equipments

☆ Mold and Deckle

☆ Blender fitted in a tank of concrete

☆ Handmade VAT

☆ Screw press

☆ Calendering press

☆ Cutter

☆ Miscelleneous like clips, rope, hammer, plastic mug, nylon mesh, rubber lining, Gada cloth etc.

Investment

The unit could be set up with an investment of Rs. 50,000– 100000

For production of handmade paper from agricultural residues, an investment of Rs. 7-10 lakhs on equipment such

Waste paper

↓

Manual shredding

↓

Soaking water over night

↓

Blending in a mixer for preparation of pulp

↓

Transfer pulp to mold and deckle

↓

Formation of sheet

↓

Press sheet in screw press to get smooth sheet

↓

Drying

↓

Cutting to size

↓

Calendering

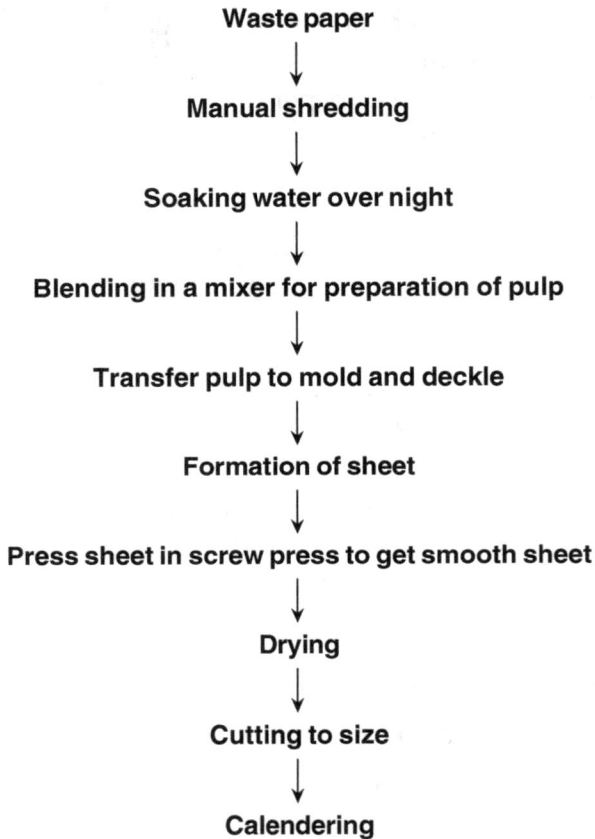

Process Flow Chart (Only from Waste Paper)

as beater, agitator, tank, cylinder mould machine, drier, calendaring machine and cutting machine is required.

Capacity   300 sheets/day

Uses

Using the sheets made through the above-mentioned process one can make the following paper products:

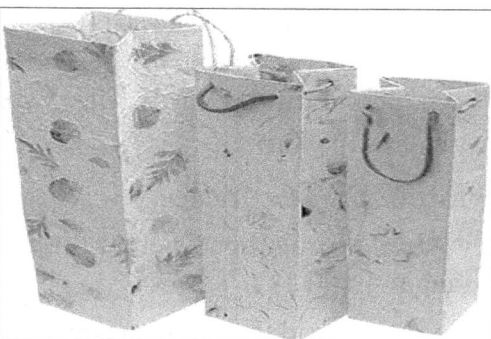

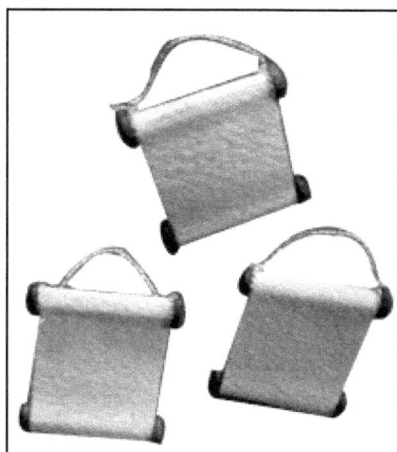

Writing papers, Tissue papers, Gift bags/box, greeting cards, Book covers, paper towels and plates, Business cards, wall papers, Birthday cards, File covers and folders (small size) and Envelops

Source of Technology

AMM Murugappa Chettiar Research Centre, Taramani, Chennai – 600 113.

Note: Abundant water supply and cheap labour should be available apart from 50 KW load. Some units are working successfully at Kalpi (near Kanpur) Uttar Pradesh. M/s Development Alternatives, Orchha (near Jhansi) MP and a unit at Itarasi, Madhya Pradesh.

29
Rice Husk Cement

In the primitive times pucca house used to be constructed using bricks and for joinery and plastering materials like lime, surkhi (brick powder), waste of jaggery production (molasses), finely ground slurry of urad dal were used. These materials were mixed in certain ratio and processed which required labour and skill.

The forts and historical buildings several hundred years old are testimony of durability and strength of these materials. Recently, it has been found that waste water obtained while cooking rice and lime were used during construction of great China wall which has good bonding strength.

Rice Husk

The housing changed with passage of time with advent of modern building materials like cement. Due to spurt in housing construction activities the cost of building materials has increased. The organizations like Central Building Research Institute, Roorkee (CBRI) and Advanced Materials and Processes Research Institute (AMPRI), Bhopal carried out Research and Development work for low cost housing and alternate building materials using agricultural wastes like rice husk (which has more than 70 per cent silica), clay, lime and industrial wastes like red mud of Aluminium Industry.

The cost of cement which is used for RCC work, joinery and plastering is increasing every day. It is high time to use cheaper but dependable substitutes. One such material is rice husk cement.

Process

☆ In the process for production of rice husk cement, rice husk, lime and clay are mixed in ratio of 50:25:25 by weight and slurry is made using water.

☆ Subsequently nodules/round balls of 5-10 cm diameter are made manually in cottage scale operation and by using noduliser in mechanized unit.

☆ The balls/nodules are sun-dried and later fired in a trench kiln. Except initial burning no fuel is required due to heat of combustion of rice husk.

☆ The sintered burnt material is micro-pulverised.

☆ The grinding process is continued for 45-50 minutes to get fine mesh 4000 per cm^2/g.

Raw materials

↓

Mixing in prescribed ratio

↓

Conversion into slurry with water

↓

Nodules/balls making

↓

Sundrying

↓

Burning in kiln

↓

Pulverisation

↓

Weighing

↓

Packaging

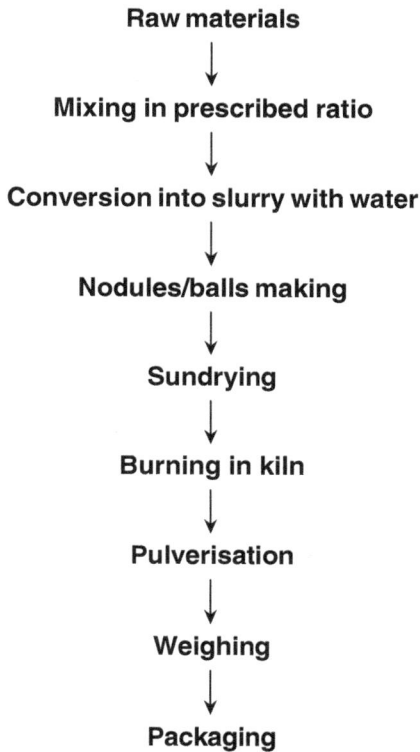

Process Flow Chart

In the AMPRI process, red mud, rice husk and coarse sand (murram) are mixed in prescribed ratio and fired in a kiln. The ash so obtained is ground with slacked lime using micro-pulverisers. The pulverized material is then weighed and packed in bags for marketing.

The rice husk cement could be used for joinery, plastering, soil stabilization, stabilized bricks etc. but not recommended for RCC/Roofing construction.

It may be noted that sludge of industry could also be used.

The hydraulic rice husk binder has a bulk density of 360 kg/m³, which increases to 700 kg/m³ on grinding.

Techno-Economics

Raw Materials

Clay	Available locally
Lime	Available locally
Rice husk	From rice mills
Red mud	From aluminium industries

Equipment and Machinery

Kiln, ball mill, noduliser, pulveriser, sieves, weighing machines and product testing equipment.

Pulveriser

Kiln

Nodulizer

Physical Characteristics

(a) Fineness= 150 micron (retention not more than 15 per cent)

(b) Std. Consistency = 38.5 per cent

(c) Setting time 1. Initial   = 2-3 hours

               2. Final   = 10-13 hours

(d) Comp. Strength

    1. At 7 days = 2-3 N/mm²

    2. At 28 days = 4-6 N/mm²

(e) Shrinkage   = 1.0mm

Costing of machinery   5-6 lakhs

Production capacity   5T/day

The cost of production varies from place to place depending upon availability of wastes like sludge, lime, rice husk etc.

The selling price will be 1/3rd of present marketing price of cement.

Source of Technology

CBRI, Roorkee and AMPRI, Bhopal.

30
Lemon/Orange Peel Oil

Waste to Wealth

Lemon (lime fruit) is consumed in every house hold in urban areas and after extraction of juice the peels are thrown into dust bin. Lemon squash industry has also abundant quantities of peels. The lemon peels contain 1.5-2.0 per cent lemon peel oil which is used in food flavour and cosmetic industries. Similarly, the orange peel oil in cosmetics, food and flavour is largely demanded industries which could be extracted from peels which is waste of food industry. The price of these range from Rs. 2,000-3,000 per litre.

Collection of lemon/orange peels could be organized through rag pickers, juice centers, food/squash Industries.

The process of manufacture involves washing of peels, separation/shredding of peels, steam distillation of peels to get essential oil of lemon peels. The equipments are required washing tanks, shredder, pomace separator, distillation column and baby boiler etc.

Those interested to invest in higher capital investment can use pomace and can produce pectin which is mainly used as gelling agent in the making of Jams, Jellies and Marmalades. Food processing industry requires pectin in process operation like clarifying, thickening, stabilizing or foaming agent. There is considerable demand of pectin in pharmaceutical industry in formulations for use in diarrhea in infants.

31

Caffeine from Tea Waste

In the processing of tea, a good amount of waste accumulates in the form of sweeping of fluff, stalks and leaves. Most of this waste is at present destroyed to prevent adulteration. It has been found that tea waste contains 2.0-3.5 per cent caffeine,which makes tea waste an attractive source of caffeine.

Caffeine is widely used in many pharmaceutical preparations as a stimulant of the central nervous system and also as diuretic. It gives definite relief from minor fatigue and neuralgia and is useful in headaches originating from eye strains. It is also used in cola type carbonated beverages.

In the process the waste is cleaned and boiled with lime liqour for about 2 hours. The mass is then filtered hot and the filtrate is cooled. Caffeine in the aqueous phase (filtrate) is extracted counter-currently with benzene. The extract containing caffeine is subsequently evaporated. Tea vapours are condensed to recover benzene for re-use. Caffeine thus obtained is dissolved in water and subjected to crystallisation to get pure caffeine.The yield is about 1.5 to 2 per cent by weight of tea leaves.

The raw materials needed are tea waste, lime, benzene and active carbon. The plant requires water, steam, fuel oil and electricity as utilities.

Equipment

☆ Vibrating screen

☆ Cooler-cum-mixer

☆ Extractors

☆ Condensors

☆ Layer separation tanks

☆ Storage tanks (Benzene etc.)

☆ Vacuum pump

☆ Circulation and other pumps

☆ Evaporators

☆ Dissolver

☆ Filtraton unit

☆ Crystallizer (Jacketted)

☆ Centrifuge

☆ Dryer (hot air circulation)

☆ Pulverizer

☆ Electrical motors

☆ Instruments, piping, hardware items etc.

☆ Boiler

☆ Water treatment plant

☆ Cooling tower

☆ Storage tanks (water, oil)

☆ Transformer

☆ Refrigeration unit

☆ Water circulation pump

Product Specificatons

The product obtained conforms to the BP specifications.

Plant Parameters

☆ Production capacity: 30 tonnes per annum

☆ Number of shifts: Three per day

☆ Investment (excluding land and buildings)

☆ Area: Total: 8000 sqm.

    Covered: 540 sqm

☆ Machinery: Rs. 40.00 lakh

☆ Manpower:

    Managerial: 06, Technical : 10, Others : 12.

The investment is around Rs. fifty lakhs.

32
Rice Husk Ash Insulating Tiles

It is entirely a new concept. In a tropical country like India the insulating tiles could be used in buildings/housing to bring down room temperature by 5-7°C and thus reduce the cost of cooling/air conditioning It could also utilize the rice husk ash which is a waste. Process involves mixing of rice husk ash (90 per cent) with 10 per cent gypsum/ Phosphogypsum (Waste of fertilizer industry) and processing the mix in the form of tiles (9" x 12" x 1½″) and finally sintering the formed of tiles in a furnace. The tiles could be used on roof tops and exteriors.

POLLUTION
CONTROL

33
Activated Carbon Mask

Exhaust of Automobiles is major cause of pollution in metros, sub-metros and even small towns. The Government has taken steps to reduce the pollution by controlling exhausts and introducing CNG in some metros. The fumes of these exhausts have obnoxious gases like Lead Oxide, SO_2, CO, CO_2 etc. which cause asthma, lung diseases, cancer etc. As such one has to take preventive measures himself.

One such option is use of activated carbon mask which absorbs these gases and allow only intake of fresh air.

One can use this mask for a week and on off days keep the mask in an oven for release of these gases into the atmosphere.

In this mask activated carbon cloth is used as media for absorption of gases.

The entrepreneurs can purchase carbon fiber from M/s. Indian Petro Chemical Corporation (IPCL), Vadodara weave the cloth and prepare mask.

The technology for production of activated carbon Mask is also available form CSIR's National Physical Laboratory, New Delhi.

34
Lead from Battery Scrap

The waste of lead - acid battery is major cause of pollution and it is very essential to dispose off scrap. It is a profitable industry to recover the pure lead from lead scrap of battery industry. The container is re-used and could also the purified. The lead scrap is a major concern. In the recovery process battery scrap is mixed with reductants and fluxes. The mixture is compressed to form briquette followed by smelting in the furnace to recover lead. The process recovers more than 90 per cent lead metal with purity level of 98 per cent. The sulphur in the battery scrap is converted to sulphide and could be used for producing H_2S. The dust emanating during the process is collected by providing dust catching arrangement.

The major raw materials are lead scrap, reductant, charcoal and flux. The equipment required includes rotary furnace with dust catching arrangement, mixer and press.

A plant of capacity 1-3 tpd. could be set up with investment in range of Rs. 5-15 lakhs.

For knowhow CSIR National Metallurgical Laboratory (NML), Jamshedpur could the contacted.

POLYMERS

35
Adhesives

Adhesives have been used since time immemorial for joining of wood, rubber, stationery/books etc. Initially raw materials used were based on natural materials like gum Arabic, maize/potato starch, dextrin. These raw materials have been replaced by synthetic chemicals to get better bonding strength and diverse uses.

Some of adhesives required in market are

☆ Rubber based contact adhesive: used in shoe, furniture, automotive and construction industries.

☆ Label adhesive: used for labeling beer and whisky bottles by automatic machines.

☆ PU adhesive: Used for permanent bonding of shoe soles.

☆ Gasket adhesives: Bonding asbestos to AL/cu.

☆ Paper to oil coated tin label: Labels on oil tin.

☆ Cyano acrylate adhesives: Bonding metals, plastics, rubber, glass.

☆ Synthetic stationery adhesives: Bonding paper to paper.

☆ Pressure sensitive adhesives: for bonding paper to paper in envelopes.

☆ PVA based adhesive: General purpose adhesive for stationery to wood.

☆ Dental/tissue adhesive: for fixing artificial teeth and join tissues after surgery.

Technology is available from Indian Institute of Chemical Technology, Hyderabad.

36
Plastic Industry

Plastic industry of Rs.85,000 crores revenue is growing more than 10 per cent and processes 10 million tons of plastic (7.5m virgin 2.5 re-cycled materials) which provide livelihood to several (08) million peoples in the country. Besides it provides market to resin products, processing machines and mould makers. 600-700 million Indians are using attractive and economical plastic retail packs and pouches, packing crispy or fresh snacks, ready to eat food or the cereals or atta (flour), the noodles or pasta, the candies or sweets, beverages, oils and creams, pickles or spices, shipping consignments, domestic furniture, corrugated sheets, walls, water tanks, temporary tenements etc. Fancy articles like ladies bags, folders, matts manufactured from waste offer opportunities for lively hood. It is very difficult to think of life without use of plastic/ polymers. There is much negative publicity about plastic waste. The rag pickers collect the plastic wastes and/supply to re–processing industries and result in 25 per cent share in production of plastic goods. Waste is also used for land fills. There is suggestion to use plastic waste 25-30 per cent mixed with bitumen in roads making. It improves the life of bituminized roads. Revolutionary materials have been developed for building construction and furniture industry made from thermo set resins and renewable natural fibers like jute, sisal etc. Plastics are also used in automobile industries, agriculture, irrigation pipes, pharmaceutical and health care products, food processing crates, telecom and IT industry (sleeves for cables) domestic items, construction industries and packaging industries, electrical conduit pipes, wires, sanitary pipes to replace GI pipes etc.

The machinery required in manufacturing of plastic goods like injection molding machines, extruders, blow moulding machines, extrusion stretch moulding machines are available.

☆ *Extrusion Process*: Rod, Blown Film, Tubes, Sheets, Profile, Blow mouldings/Cables, Laminates.

☆ *Extrusion of Blown film*: Sacks, bags, sheets and Protecton cover for all type of goods for its transparency.

☆ *Extrusion coating*: Coating of very fine film to improve impereability and weldability.

Use in packaging food, milk pouches, frozen foods cakes, meat, Pharmacuiticals, Paper cups/plates.

☆ *Co-Extrusion*: Multi layered plastic articals and laminations.

Thermo formed Trays, cups.

☆ *Pipe - Extrusion*: Agriculture, house wiring. Water tap, Chemicals, Sewage pipes.

☆ *Injection Moulding*: Grannuls are heated, plasticisers are admixed and material is passed into mould and pressed by using plunger.

☆ *Blow Moulding*: Use air blow to shape plastic material. Bottles, Hollow objects.

Tubes are introduced into mould in soft condition and air pressure is applied, then allowed to cool, air pressure is dropped and item is ejectd out.

☆ *Stretch Blow Moulding*: Rigid plastic containers for edible oil.

Beer and Carbonated soft drink, Mineral water, Cosmetics and Toiletries etc.

Machines are indigenously available and quality of machines is comparable to any imported machine. The cost of machines depend upon the product selected for production ranging from Rs. 5 lakhes to few corers. The raw material like PVC, polystyrene, polythene, polyolefin, polycarbonates, polypropylene, high density polyethelene are available in the country. Range of plasticizers needed for processing are also available indigenously. The entrepreneurs can take technology/guidance from Central Institute of Plastic and Engineering Technology (CIPET), Chennai or through its centers spread over the country.

Plastic consumption in India is set to increase from 7kg/ person/annum to 26kg/person/annum in the next ten years.

37
Smart Cards

There is boom for Smart Card, RFID and Biometric Industries in India. Smart card, RFID application are currently proliferating into newer socio-economic areas. India has now emerged as the world's biggest market of bio metric applications and e-payments and Mobile payments are the current buzzwords, proving the most effective vehicle for financial transaction

The automated Finger print identification system and UID Biometric implementation are taking now strides. Indian Biometric scenario has growing opportunities.

Source of Technology/Guidance.

1. CDAC, Mumbai.
2. Bharat Technical Solution Pvt. Ltd.
3. National Informatics Centre (NIC) New Delhi
4. Indian Institute of Technology (IITK).

EMPLOYMENT GENERATING TECHNOLOGIES

38
Neem Oil

Neem has been in use since time immemorial in daily habits of Indian village folk. Leaves were used for preservation of grain, safety of cloth from insects etc. At present very large quantities of seeds go waste. Collection of neem seeds can provide jobs to very large section of rural society for at least 70 days during April-June. Neem seeds contain, about 10 per cent neem oil which has demand in soap industry, water emulsion neem oil as pesticide. Besides neem oil has good export potential. De-oiled neem cake is used in agriculture fields as biomanure and also kill harmful bacteria.

The process involves decortication of seeds, removal of yellow skin, washing in water, (ii) Crushing of seed in Jaw crusher (iii) extraction of oil by power driven ghani rotaries and expellers Use modified expeller developed by CSIR

Micro-Enterprises in Agriculture

CMERI Centre, Ludhiana Punjab, and lastly (iv) Filtration of oil.

1)	Capacity	1 T of seeds/day
2)	Building	2000/sq.ft. Rs. 3.00 lakhs
3)	Machinery	
	Scrubbings and washing unit	
	Jaw crusher	Rs. 4.00 lakhs
	Expeller	
	Filter press	
		Rs. 7.00 Lakh
Working capital		Rs. 1.00

Note: The machinery could be used for extraction of oil of other oil seeds like Sunflower, Mustard, Groundnut, linseed etc.

The liquid obtained during washing of seeds could be used as pulp for feeding Honey Bee when flora and fauna is dry.

39

Herbal Phenyl

NEEM is widely available throughout the country as wild growth or planted in villages/suburbs for shade. The pesticidal and insecticidal properties of Azadiric-achitin present in Neem have been internationally acclaimed now, although Indian homes using neem leaves from time immemorial for preservation of food grains and woollen clothes.

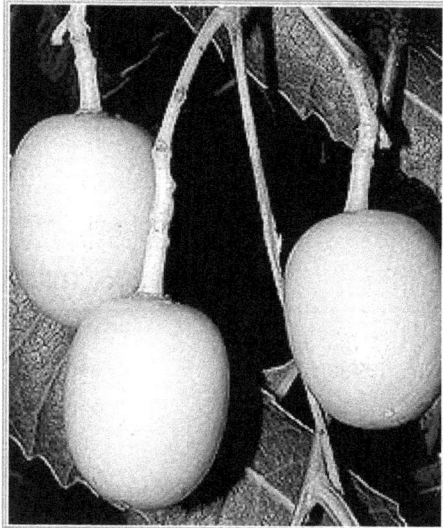

Micro-Enterprises in Agriculture

Now a days great stress is given on use of herbal products in the country as well as all over the world due to absence of any side effects.

Applications

☆ It is suggested to manufacture Neem Phenyl for use in houses, hospital etc. in place of chemical phenyl.

☆ The product would find wide market.

☆ The ingredients used in this formulation would also repel flies, mosquitoes and cockroaches.

Raw Material

Following raw materials are required in the manufacture of Herbal Phenyl which is easily available in the local market.

Sl.No.	Material	Rate
1.	Neem Oil	Rs. 100/litre
2.	Water	-
3.	Citronella Oil	Rs. 700/litre
4.	Colour	Rs. 700/kg

Material Mix for 100 litre of Phenyl

1.	Water	90 per cent	90 litre	Rs. 2.00
2.	Neem Oil	5 per cent	5 litre	Rs. 500.00
3.	Citronella Oil	1 per cent	1 litre	Rs. 700.00
4.	Colour	-	200 ml	Rs. 140.00
5.	Packaging materials	-	-	Rs. 1000.00
			Total	Rs. 2340.00

Say Rs. 2350/- for 100 litres

Machinery Required

Sl.No.	Machinery	Cost
1	Mixer-cum Emulsifier of 200 litre capacity	Rs. 25,000
2	Packaging machine	Rs. 10,000
3	Measuring cups and misc.	Rs. 5,000

Process of Manufacturing

Raw materials-water, neem oil and citronella oil are mixed in proper proportion and transferred to emulsifier. The emulsification process is carried out for 4-6 hr. the emulsified material is filled in PVC bottles/cans and sealed by using cap sealing machine.

Market Demand

It is a herbal phenyl that can replace all the other phenyls available in the market. It not only cleans the floor but also acts as a repellent for cockroaches, mosquitoes, flies etc. It disinfects also. Hospitals, nursing homes, offices and homes form a big market of herbal phenyl.

For further information entrepreneurs can contact the author.

40
Rabbit Farming

There is wide spread unemployment in rural and tribal areas of country more than 55 per cent population is living below poverty line, as such, it is very essential to promote newer means of employment and generate avenues for augmentation of income by utilising idle time. The female labour force participation is very low. It has been found that animal husbandry plays very impotant role in mitigating poverty among landless labour and marginal farmers and the activities like poultry and rabbit farming could be better performed by women.

Rabbit farming is profitable enterprise in rural and tribal areas with low initial investment. It is now being persued in cold climatic zones of country like Himachal Pradesh, Uttarakhand, Jammu and Kashmir, north-eastern states and hilly areas of South India. There are two varieties of rabbits adopted for Commercial use *i.e.,*

1. Angora for wool and fur,

2. Chinchilla for meat.

Angora wool has a good demand in Indian woollen industry. The price of angora wool is not less than Rs. 2000/- per kg.

The angora rabbits are reared in low temperature zone falling in the temperature range of 5-25°C.

Chinchilla variety is used for meat production. Rabbit meat is almost zero cholestrol meat and these could be reared at places of temperature upto 35°C. Artificial cooling is provided during summer months.

Production and Breeding

Rabbit is a fast multiplying small animal. A starting unit comprises two males and 10 females. After one year the animal populaion is in the range of 150-200, mating age is 6-8 weeks and pregnancy period is 30-35 days and a female gives a litter comprising 7-8 offsprings every 3-4 months.

Feed Management

Rabbit is herbivorous animal and eats green feed, soft hay grass and kitchen green vegetable waste in addition to 15 per cent pelletized solid feed. If pelletised feed is not available take 'makka' and 'jowar' and coarse grind and mix. Feed requirement is 3.7-4 per cent of body weight (approx 4.5-5.0 kg per animal) which is 150-200 gm and energy requirement is 1000-1200 cal/kg of body weight.

Housing Management

The rabbits are housed in cages of size 24"x18"x18" kept in 2-3 tier racks. The cages should not have sharp edges otherwise rabbits would be hurt. The breeding is done at places away from population, preferably in green areas free of noise, smoke, fumes and protected from predators like dogs, cats, jackals etc.

Health Management

Diseases lika scabies, ear mange, weepy eyes, bronchitis and diarrhea are common among rabbits. Proper saniatation and feeding minimise diseases. Common medicines suitable for human beings are administered

Do's and Don'ts

☆ Handle rabbits gently avoiding pressing and jerks of their body.

☆ Regular clearing of feed and water bowels.

☆ Provide sufficient water.

☆ Mating should be done during early hours of morning and female is brought to male.

☆ Check the belly of female after 15 days by gentle pressing.

☆ After 25 days bed should be prepared for pregnant does.

☆ Keep animals in separate cages in cool climate.

Production

Shearing of wool (2-3 inch long) should be avoided prior to winter months but should be completed on the onset of summer.

Marketing

The pooled marketing of a cluster should be done to meet the requirement of bulk purchasers. Help of some NGOs could be sought in early years. Large scale units do their own marketing. They could also be approach woollen mills.

Sale of fur for handicrafts like caps, coats, bags, purses, toys, dolls would lead to additional revenues.

Techno-economics (150 nos keeps 10 per cent mutually)

Cost of 10 females + 2 males @ Rs. 500 per rabbit	Rs. 6000
Cost of cages 150x200 nos.	Rs. 30,000
Feed for one year	Rs. 45,000
Medicines and labour (self + one help)	Rs. 30,000
	Rs. 1,11,000
Cost of production	Rs. 1,10,000
Interest	Rs. 10,000
	Rs. 1,21,000
Sales 100 kg of wool @ 2000/-/kg	Rs. 2,00,000
Sales fur	Rs. 10,000
Net sales	Rs. 2,10,000
Cost of production	Rs. 1,20,000
Profit	Rs. 90,000
Profit	Rs. 7,500/month

Source of training and rabbits

☆ Sheep and Wool Research Centre, I.C.A.R., Garsa, Kullu, Himachal Pradesh

☆ Central Avian Research Institute, Izzatnagar, Bareilly, U.P.

41
Micro-Concrete
Roofing Tiles

Housing is the basic need of every human being. Central Government as well as State Governments are keen to provide houses to everyone and especially to weaker

sections of the society. The cost of housing construction is increasing day by day due to rising cost of building materials. The government sponsored agencies and individuals are looking towards low-cost building materials without compromising with the quality. Research and Development organizations of the country have developed low-cost building components for walling and roofing which enable one to build improved houses at comparatively cheaper cost. Price of corrugated G.I. sheets used for roofing, is continually increasing while its use does not provide thermal comfort in a tropical country like India. Use of Asbestos-cement sheets is being discouraged due to health hazards associated with asbestos fibre. MCR (micro concrete roofing) technology meets the growing demand for high quality roofing. MCR tiles are cost effective and extremely versatile roofing material. MCR tiles can be used on steel and wooden under structure to make attractive roofs on residences, farmhouses, *verandahs* and pavilions.

Advantages

MCR tiles have several advantages:

1. These are water and fire proof
2. These are eco-friendly
3. Cheap in cost
4. Utilises waste (stonedust)

Process of Manufacturing

1. Step I

☆ Selection of raw material

☆ Cement (43 grade)

☆ Stonedust (fine)

☆ Aggregate (<6mm)

☆ Water

2. Step II

Mix Preperation

The above raw materials are mixed in the following ratio:

Cement	:	Stonedust	:	Aggregate	:	Water
1	:	1.5	:	1.5	:	0.5
						(by weight)

3. Step III

Mix Vibration

The prepared mix is put on a polythene sheets and placed over vibration table for about 45 sec. a nail is inserted in the rib for holes.

4. Step IV

Moulding and Stock Curing

The polythene sheet is then carefully placed on the plastic moulds for getting a uniform shape. The moulds are then stacked one above the other in a set of 25 for next 24 hr.

5. Step V

Demoulding and Water Curing

The tiles are then demoulded and seperated from the plastic sheets. The nails are also taken out and edges sharpened for better finishing. Then the tiles are kept in the curing tank for another 7 days.

6. Step VI

Strength and Quality Testing

After the tiles are taken out from the tank these are stored for next 21 days in open or shed. Tiles from the stock are taken at random for testing in accordance with standard specifications. Then the tiles are ready for sale.

Product Specifications

Clear length	488 mm
Length after overlap	400 mm
Clear width	240 mm
Width after overlap	200 mm
Thickness	8mm/100mm
Corrugation depth	50mm
Normal weight	2.25-2.60 kg
Purlin spacing	400 mm
Load bearing capacity	80/100 kg

Selection of raw materials

↓

Mixing of raw materials

↓

Vibration of mix on vibration table

↓

Moulding

↓

Demoulding

↓

Water curing

↓

Drying

↓

Random testing

Process Flow Chart

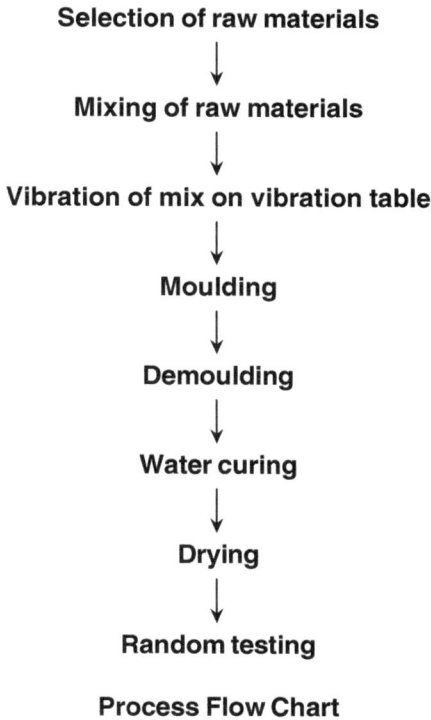

For covering 10 sq. m. (100 sq. ft.) roof area 130 tiles are required.

Operating Parameters

☆ Capacity: 200 tiles/day

☆ Machine/equipment:

Ø TARA-MCR Machine: 1

Ø Moulds (PAN): 200

Ø Alignment tool: 200

Ø Bend test equipment: 1

☆ Power requirement: 5 units/day, single phase 1 (one) KW electric connection would be sufficient.

✰ Other requirement: water 100 L./day

✰ Raw material

Ø Cement: 2 bags/day

Ø Sand/stone dust: 4 qt./day

✰ Packaging: No special packaging is required

✰ Manpower requirement for production: 4

Techno-Economics

a) Land and building with auxiliaris, 300 Sq. ft. covered area: Rs. 1,00,000

b) Machinery and equipment: Rs. 1,00,000

c) Working Capital: Rs. 2,00,000

d) per cent Profit: 30 per cent

Selling price per tile: Rs. 12.00

M/s Development Alternatives a NGO of repute has developed the technology and machine manufacturer of MCR tiles.

Entrepreneurs can contact state rural engineering services for supply of tiles.

42
Indigo Blue Dye

Indigo blue was rare commodity in Europe during the 19th and 20th century and to meet its demand the British coerced farmers in India to cultivate planting material for indigo production. In 1987, 7000 sq km area was under indigo cultivation in India. The play *Nildarpana* by Desh Bandhu Mitra is based on indigo slavery and forced cultivation of indigo in India. Subsequently, indigo was produced by using synthetic chemicals.

Now the trend all over the world is to use natural dyes and colours which may revive this organic dye. Indigo is used in textile mills, laundries and households to give bluish tinge to white cotton clothes. Other natural fibres like velvet silk, munga silks are dyed using indigo blue. It has no harmful effects and toxicity to skin. Indigo dye is used for denim cloth for jeans, wool and silk. When ironed the cloth gives a shining finish. In different parts of the country people follow their traditional methods for preparing natural dyes. The realization of hazards caused by synthetic dyes has made the society revert back to traditional raw materials.

Indigo which is among the oldest natural dyes has been used for textile dyeing and printing. India has also been one of the earliest major centre for its processing and production. The *Indigofera tinctoria* variety of Indigo was domesticated in India. It then made its way to the Greeks and Romans who valued it as a luxury product. India was a primary supplier of indigo to Europe as early as the Greco-Roman era.

In the modern era Shri AMM Murugappa Chettiar Research Centre (MCRC), Chennai, has been involved in extracting dyes from natural sources for more than a decade. The dye pigments from bacteria, indigo blue dye from *Indigofera tictoria, Writtia tinctoria,* and *Tephrosia sp.,* red and orange dyes from *Morinda tinctoria* and *Bixa orellana* respectively and brown dye from *Acacia catechu* are the major scientific achievements of the Centre.

For blue natural dye the only viable choice is indigo from plants. The plant is grown in Asia, Africa, East Indies, Philippines and America. In MCRC process natural indigo is obtained by fermenting the foliage of species of Indigofera and a good tech-pack is now available for use. All the processes involved in this package of practice are eco friendly and cost-effective.

The extraction of indigo from the plant consists of mainly 3 stages *viz.* steeping the plant in water (for fermentation), separation of the aqueous extract and oxidation of this solution with air and separating the precipitate and preparation of marketable dye cake or powder. MCRC has developed a new microbial process which has been found efficient for extraction of indigo dye. A patent has been filed on the technology package. Under CAPART's project more than 100 various Indian organizations have been trained on natural dyes, especially on indigo.

MCRC's Indigo Extraction Technology Package Consists of:

Organic Plant Production

Cultivation methods for indigo plant includes crop maintenance and crop protection by biological control and use of residual green matter as green manure and for compost preparation.

Dye Extraction Techniques

Harvest of plants, microbial fermentation of plants, culturing selected bacteria by simple method, use of netlon soaking set-up, mechanical oxidation, slurry collection, boiling and filtering.

Technical Know-how

Construction and design of indigo tank (MCRC Model), fabrication and maintenance of indigo agitator and fabrication of netlon soaking set-up.

Aspects of Indigo Market

Indigo plant, dry leaf, indigo dye and residual matter are sold in the market. For MCRC's indigo dye extraction, initial investment of around Rs. 60,000/- for a unit is required. Minimum 15 acres of land are required to support an extraction unit for 200 working days in a year.

Economics

Cost of cultivation	Rs. 5000/per acre
Biomass obtained	4-6 tonnes
Return	5000 kg @ 2/- kg= Rs. 10,000 per acre
Processing cost	Rs. 1/- per kg
Total cost of production Rs. (2+1)	Rs. 15000/-
Sale value received @ Rs. 4/- per kg or Rs. 400 per kg of indigo	Rs. 20,000/-
Sale of residual green matter @ Rs. 0.50 × 5000 kg	2,500/-
Total sales	22,500/-
Profit	22,500-15,000 = Rs. 7,500/- per acre
Profit per year from 15 acres 7500×15	Rs. 1,12,500/-

Source of Technology

Shri AMM Murugappa Chettiar Research Centre (MCRC), Chennai

43
Ferro-cement
Water Tanks

The word "Ferro-cement" was coined due to use of Ferro (Iron) and Cement as the main raw materials. Ferro-cement tanks are lighter in weight (wall thickness of 500 liter is one cm) and higher in strength as compared to re-inforced concrete tanks. It involves minimum skill, lesser project cost for fabrication, maximum utility and service-ability. Ferro-cement tanks can replace costly steel/mild steel tanks. As compared to PVC and steel tanks, ferro-cement tanks are eco-friendly and the quality of these tanks is that here the water gets less heated in summer.

Ferro-cement technology has diversified applications in addition to water tanks such as grain storage silos, septic tanks, animal feed bins, cup-boards, boundary wall, door shutters, segmented roofing sheets, man-hole covers (light,

medium and high duty), irrigation channels, ferro-cement earthquake resistant houses, flower pots etc. All these products can be fabricated in the same unit with the use of different moulds.

While tanks upto 1000 L capacity could be fabricated at factory premises higher capacity tank could be fabricated by casting the cylindrical surface in four segments which are later joined together with base *in situ*. These segments are transported in trucks carrying the material for large number of tanks.

Water tanks of capacity ranging from 200 L to 10,000 L are in great demand for storage of water due to the problem of water scarcity. These are being used in residential houses, governmental organizations such as housing boards, slum clearance boards, Public Health Engineering Departments, municipal corporations, rural community water supply etc. CSIR Labs have transferred this technology to more than 150 parties and ferro-cement tanks are in use in various parts of the country.

Raw Materials

High tensile galvanized wire mesh, mild steel bars, cement, sand, water and food grade resin are the main raw materials.

Plant and Machinery

1. 3/2 ft concrete mixing mechanism
2. Welding transformer
3. Bar bending benches, levers
4. Moulds, masonary tools

Civil Work

1. Shed of 20'×20'
2. Curing tank
3. Casting yard

In the Ferro-cement techniques high strength galvanized wire-mesh is sandwiched between layers of cement and sand mortar in the ratio of 1:2.

Ferro-cement water tanks have replaced M.S. and concrete tanks due to cheaper cost, lightweight and long life requiring almost zero maintenance. There is good demand of water tanks in urban as well as rural areas.

In a semi-mechanised process wire mesh is wound tightly around cylindrical mould and cement sand mortar (ration 1:2) is applied in such a way that wire mesh is sandwiched between the layers of mortar so as to achieve wall thickness of 12 mm for tanks upto 500 L capacity. Base and cover are casted separately. Base is joined to the cylindrical surface. The tanks are water cured for 15-20 days by sprinkling water or immersing in water. The inside surface of the tank is coated with anti-fungal/anti algal food grade resin. Outer surface is painted with silver paint to reflect heat.

Investment

Production capacity of 10 tanks of 500 L capacity requires investment of Rs. 3.50 lakhs.

Fixed Capital

Equipment and moulds	1.50 lakh
Shed 20'x20'	1.50 lakh
Utilities and Miscellaneous	0.50 lakhs
Total	3.50 lakhs
Selling Price of 10 tanks per day @ Rs 1500/tank	15000/day
Cost of Production	Rs. 10,000
Profit	5,000/day

The following products also can be manufactured in the same production facility with minor addition of moulds:

☆ Ferro-cement grain storage silos/bins.

☆ Ferro-cement door shutters.

☆ Ferro-cement roofing sheets.

☆ Ferro-cement manhole covers.

☆ Ferro-cement irrigation and drainage channels for rain water harvesting.

☆ Ferro-cement earthquake resistant house.

☆ Ferro-cement cupboards.

☆ Ferro-cement septic tanks.

☆ Ferro-cement flower pots.

☆ Ferro-cement animal feed units

☆ Ferro-cement tree guards

Entrepreneurs interested in technology transfer and training may contact the author or CBRI, Roorkee, SERC, Chennai.

44
Lead-Acid Battery

Battery industry is progressing at very fast pace. Starting from simple lead-acid batteries, Lithium ion battery. Now research heading towards nano-carbon tube batteries which would store very large energy in small size. Lead - Acid batteries find variety of applications such for motors, cars, two wheelers, tractors, trucks, air craft, submarines, stationery batteries for telephone exchange, deep discharge batteries for inverters etc. Automobiles is the largest user of batteries. The main purpsose of using these in automobiles is for ignition and lighting. CSIR's Central Electrochemical Research Institute (CECRI), Karaikudi (TN) have developed a wide range of batteries for diverse applications. The CECRI provides training and technology using indigenously available raw materials covering testing and analysis of raw materials, quality control procedure and testing of finished products.

Selection and testing of raw materials

↓

Grid Casting

↓

Mixing, Pasting and curing

↓

Forming and drying

↓

Assembly

↓

Testing

Process Flow Sheet

The unique feature of technology is formulation of paste with incorporation of suitable additives for the production of positive and negative plates. The other components such as hard rubber containers, covers, separators procured from standard suppliers. The know how is well suited for small/micro level entrepreneurs to produce product comparable to large scale units confirming to BIS specifications.

Raw materials required are such as lead-antimony alloy, lead oxide, sulphuric acid, barium sulphate, carbon black, containers with cover, separators etc.

Equipment and Machinery

☆ Alloy melting furnace

☆ Grid moulds

☆ Mechanical mixers

☆ Lead welding equipment

☆ Acid storage tanks

☆ Distilled water plant

☆ Rectifiers

☆ Battery chargers

The know how is proven and is being used by large number of SSIs and Companies like HAL Bangalore.

For a small plant of capacity 20 batteries per day (6000 p.a.) (12 V, 13 plates, 90 AH battery) investment of Rs. 35-40 lakh's required.

The interested entrepreneurs who desire to set up unit or improve their existing production techniques/facilities should contain CECRI, Karaikudi, TN.

45
Lead-Acid Storage Battery Repairing

The storage battery is a vital component for automobile vehicles. It is used for starting of the engine of cars, Buses, Trucks, Tractors, Jeeps etc. It is also used with generating sets.

There is demand for storage batteries in rural areas for Jeeps, Motor cycles, trucks, tractors etc. Battery has a certain operating life after which it has to be charged or repaired for use again.

Production Capacity

☆ Recharging: 500 Nos./month

☆ Repairing and recharging: 250 Nos./month

Process

1. Recharging - The level and density of the distilled water/acid is checked and required quantity of water and acid are filled. The battery is then connected to battery charger and charged to required level.

2. Repairing and complete charging-The damaged cells of the battery are removed and replaced with new plates and separators. The battery is filled with distilled water and acid and sealed with sealing compound. The complete charging is done and tested for voltage and current ratings.

Machinery and Equipment:

(i)	Battery chargers 4 Nos.@Rs. 5000 each	Rs. 20,000
(ii)	Soldering equipment 4 sets	Rs. 5,000
(iii)	Battery terminal machinery Equipment with blowers.	Rs. 10,000
(iv)	Misc. tool	Rs. 5,000
	Sub Total	Rs. 40,000
(v)	Testing equipment and hydrometer voltage tester	Rs. 5,000
	Total (Say)	Rs. 45,000
	Working Capital	Rs. 1,00,000

CSIR's Central Electro -Chemical Research Institute, conducts short term training programs and new entrepreneurs can contact them.

46

Medical Instruments

Stethescope invented more than century ago was an asset to check the patient and it is still in vogue for first hand checking of patients inspite of development in Science and Technology. Now a days, medical practitioners are highly depend upon diagnostic tests of patients before prescribing medicines. Medical instruments are now part and parcel for diagnosis of patients. The market demand of medical instruments is growing at very fast rate.

Keeping in view the requirements of medical world. Several R&D Institutes like CSIR's Central Scientific Instruments Organization (CSIO) Chandigarh, Indian Institute of Technology (IITs), DRDO labs, Institute of Nuclear Medicine and Allied Sciences, Delhi and National Research Development Corporation (NRDC) New Delhi have carried out R and D and developed several instruments and are in position to Transfer of Technology. Entrepreneurs should contact these organizations and have tie-up with these institutes for starting their venture.

The basic requirement is few measuring instruments for assembly. The component like LSI, VLSI, Chips are procured from market. Some of these Instruments are in good demand in Nursing Homes, Hospitals which is given below.

1. Gucometer for blood sugar.
2. Linear Accelerator for treatment of cancer patients.
3. Anaesthesia Ventilator
4. Surgical microscope
5. Dialysis machine
6. Portable E.C.G.
7. Breath Analyzer to detect Alcohol
8. HIV Kit
9. Pregnancy Kit
10. Patient monitoring system
11. Blood Pressure measuring instruments.
12. Endoscope
13. Ultrasound
14. Colour Doppler
15. Electro cardiograph
16. Urine Analysis strip
17. Digital Thermo meters
18. Cryoprobes for eye surgery.

Servicing and Maintenance of Medical Instruments

There is an acute problem of servicing and maintenance of medical instruments due to lack of specialised instrumentation services and equipment worth Crores of Rupees is lying idle in Government hospitals and patients have to suffer due to non-working of diagnostic instruments. The instruments used in Universities and Colleges of higher learning are in the same states having the same fate. Moreover, private nursing homes also need these services on priority. More than, two decades back CSIO Chandigarh were providing instrument repair and maintenance services through its 10 regional centres but these were closed due to change in R and D priorities. Entrepreneurs having specialized expertise in the field of electronics and instrumentation can set up specialized service centers in about 100 major cities of India to cater to the needs of medical world. Perhaps help could be sought from CSIO where vast R and D knowledge in the field of medical instruments sector is available.

47
Aloe Vera Juice and Gel

The history of Aloe vera is more than 2000 years old and Alexander the Great had special interest in it to heal the wounds of his soldiers. There is mention about it's use to heal wounds during *Mahabharat* due to its antiseptic and healing properties. Cleopatra used it as a skin care product and Bible mentions about use of Aloe. Aloe-has its origin in African rule continent from where it was brought to Caribbean island, India, Venezuela and Mexico. It is being cultivated in USA and its cultivation on commercial scale started in last decade. There are about 250-300 species of Aloe but few are being used. The Aloe gel contains a treasure of nutritional healing photogenic compounds. The plant contains over 250-bioactive ingredients naturally balanced in most efficient way. It contains 12 vitamins, 20 minerals, 18 amino-acids, 16 enzymes, mono and poly-saccharides, Anthraquinones and their derivatives, lignin, saponins, steroids, etc. Gel contains 99.3 per cent water and 0.7 per cent of ingredients.

The health care properties of Aloe vera are diverse and some of these are :

☆ Natural purifier.

☆ Detoxification agent-removes toxins from body.

☆ Promotes cell division helpful to patients suffering for HIV, cancer pain, asthama, skin problems.

☆ Antibacterial, anti-viral and anti-fungal properties.

☆ Skin moistures

☆ Anti-pyretic agent.

It rejuvenates the human body and slows the process of aging. The products made from juice, gel and powder have export potential to countries like Europe, Russia, America.

The machinery for processing Aloe-vera leaves is available indigenously. The machinery such as leaf washer, sorter cum cutting unit. Leaf conveyor, pulp extractor, homogenisers, special vacuum filter, hot water/steam boiler, special juice mixer, special micro filters and packaging machine, are required. The entrepreneurs can contact growers for processing leaves and marketing gel juice etc. they can contact cosmetic and pharmaceutical Industries. They can also launch their products in the market.

48

Radiator Coolant

The automotive industry is growing at a very fast rate and India is one of the largest manufacturer of four wheelers.

The coolant is used for cooling radiators and there is ever growing demand and entrepreneurs can contact service garages for supply of coolants. The coolant developed by NML Jamshedpur has anti-corrosion additives which does not corrode AL, MS, Cu, Brass, Solder, Cast Iron etc.

The raw materials required are Polyalcohal, phosphate and alkalies.

The major equipment include, tanks, stirrers, Filtration assembly, vacuum pump and weighing machine etc. which are easily available.

The plant of 200 L/day capacity costs around Rs. 10.00 lakhs.

ORGANIC LEATHER

49
Vegetable Tanning of Leather

In India Rural Leather Tanning has played prominent role for many centuries since it has benefited rural flayers, footwear and leather product manufacturing by village artisans from the tanned leathers obtained from village tanners. They cater to the day-to-day requirements of farmers and others residing in villages, and small towns. The enterprise helps substantially in the rural economy.

Rural tanners, besides procuring hides and skins from native flayers, used to collect barks, shrubs, and fruits for vegetable tanning either free of cost from nearby forest areas, trees near rivers or ponds or by paying nominal charges to its suppliers and lime was also obtained from local kilns. Thus rural tanning is fully integrated and self-sustainable enterprise.

The system suffered some setbacks as the finished leather used to emit unpleasant odour, and became hard due to incomplete tanning and did not fulfill the needs of sophisticated customers. Moreover, the tools and techniques became old and full of drudgery and forced out many artisans from the profession.

This necessitated development of improved processes for rural tanning through constant endeavors of institutions like Central Leather Research Institute, Chennai and its Regional Extension centers located in different zones of India and many State Government organizations, NGOs and Khadi and Village Industries Institutions.

It is worth mentioning that vegetable tanning is eco-friendly, discharging minimum pollutants and waste products such as bark, nuts, fruits which are a good source of organic manure for use in agriculture.

An improved and modified process of rural vegetable tanning developed at Central Leather Research Institute and demonstrated at many rural leather clusters is given below:

Raw Material

Locally procured green, semi-dried or salted hides and skins are used for vegetable tanning. Generally, buffalo and

Soaking

↓

Liming

↓

Reliming

↓

Deliming

↓

Pretanning

↓

Malani

↓

Bag tanning

↓

Setting

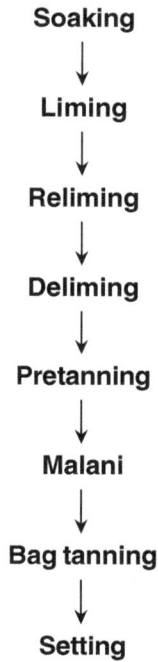

Process Flow Chart

cattle hides are processed into tanned leather. Lower grade of hides and skins are preferred and good quality hides are sold to hide and skin traders from cities and towns as they fetch remunerative prices and lower grade hides give good returns after conversion into vegetable tanned leather. Skins of sheep and goats are commonly not processed by rural tanners since their demand for manufacturing leather products is not attractive. However, in rural areas around Kolhapur and Sholapur in Maharashtra vegetable tanned goat skins have good demand for making *Kolhapuri Chappals*.

Soaking

Hides and skins are soaked either in pit of dimensions 6'x4'x4' constructed above the ground with suitable outlet of waste water into an underground drain.

While green (fresh) hides require three to four hours time for soaking but in case of salted hides soaking is carried overnight till hides become soft and pliable. Dried hides in most cases are soaked for 18 to 24 hours depending upon the condition of hides. Prior to commencement of soaking, hides and skins are weighed in order to work out the requirement of chemicals and tanning materials. However, due to any reason if hides are not weighed prior to soaking this should be followed after soaking and draining excess water from hides to record soaked weight as it helps in determining the quantity of chemicals and tanning materials in the subsequent process.

Nearly 500 grams of bleaching powder and 500 grams of washing soda (sodium carbonate powder) may be used for a pack of 10 to 20 dried hides to prevent any purification during soaking as increase in time at this juncture leads to bacterial damage. It is better to wash hides in fresh water prior to actual soaking to remove adhering dirt, blood, and other impurities from hides. Sometimes, one more wash is given after first soaking for about 2 to 3 hours so as to ensure sufficient soaking followed by liming. It may be noted that proper soaking and liming determines the quantity of vegetable tanned leathers specially from rural leather clusters.

As far as possible, hides should be flat and quite soft and slightly plumped after soaking.

Liming

Bigger hides are cut into equal half from head to tail along backbone line for ease of processing and handling by local artisans.

The soaked hides or skins are immersed in first pit containing a solution of 200 per cent fresh water and lime powder (calcium hydroxide) and 1 per cent sodium sulphide. The dimensions of lining pits are same as that of soaking pit. The liming period normally ranges from 6 to 7 days and it can be curtailed by one day depending upon the condition of soaked hides.

Hides are hauled and replaced from first pit also called old liming pit daily two times in the forenoon and evening and hauling outside for about 30 minutes. This hauling and replacing helps in agitating lime liquor and early loosing of layer from grain side and developing mild swelling and plumping. After keeping in the old lime liquor pit hides are unhaired with a curved semi blunt knife to scrap of the hair and epidermal layer from the grain surface. Hides are now taken for reliming.

Reliming

Duration 3 days

Following ingredients are added to the pit water (sufficient water to keep the hides well immersed in lime solution avoiding any sort of oxidation)

Lime powder	6-8 per cent
Sodium Sulphide	1 per cent
Caustic Soda(Sodium Hydroxide)	0.5 per cent

Similar to old lime liquor, here too hides/skins are hauled twice daily for 3 days. During this period the flesh swells sufficiently and hides are ready for fleshing.

Hides on seventh day morning are lightly rinsed in the fresh water and taken for fleshing.

For efficient and quality production it is preferable that hides now called pelt are fleshed on a wooden beam kept in inclined position. Fleshing is done using a semi curved sharp single edge knife. Some rural tanners prefer to use RAMP for fleshing and carrying out fleshing on a flat big smooth stone slab but it results poor output and causes back pain to artisans due to excessive bending of body.

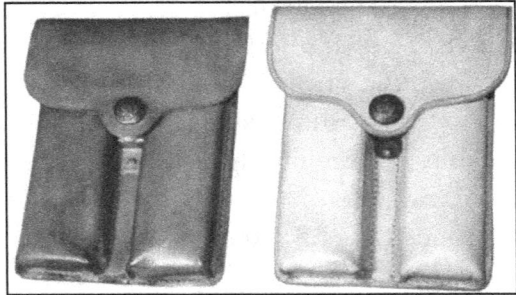

Rubber gloves must be used during fleshing work to avoid contact of hand and palm with limed pelt containing injurious chemicals like sodium sulphide and caustic soda. After fleshing pelts are scudded on grain side with semi curved blunt knife to remove hair roots, fat, etc.

Fleshed weight of pelts is noted down in order to calculate the quantity of chemicals in subsequent operations.

Fleshed pelts are washed in fresh soft water to remove surface lime, sulphide, scud etc.

Deliming

Deliming is carried out in a pit for overnight using following chemicals:

Water	100 per cent
Ammonium sulphate	1 per cent
Sodium Bisulphite	0.5 per cent

Next day pelts are trampled in deliming bath for about one hour with feet to remove the adhered lime.

Completion of deliming can be checked with phenophthalein solution. Pour few drops of phenophthalein on cut section of pelt. If it shows a streak of pink colour it indicates that deliming is almost complete. If phenophthalein is not available then deliming may also be checked with Turmeric powder solution. Pour few drops on cut section of pelt. A streak of pink colour indicates completion of deliming. After deliming pelts are once again scudded with blunt semi curved knife by placing pelt on inclined wooden plank, tree or locally available knot free beam. This is done to remove the remaining scud from the pelt.

Pelts are washed again in fresh soft water and water is drained out for 8th day.

The deliming pelts are transferred one by one to the salt solution made earlier in a pit and kept here for about an hour.

Common salt	5 to 6 per cent
Water	50 to 60 per cent

Later diluted sulphuric acid solution prepared earlier is added very slowly to the pit containing salt solution.

| Sulphuric Acid | 0.5 per cent | Pelts kept in the pit for |
| Water | 10 per cent | about 2 hours |

Note: Percentage of chemicals used above are based on pelt weight.

After further handling the pelts in the pit, pH of the cross section of a piece of pelt cut from pelt is checked with the help of pH paper and when pH reaches to 5.5 (Approx). Pretanning is carried out in the same pit so as to facilitate uniform tanning at later stage.

Pretanning

1 per cent of pretanning syntan such as Basyntan P of BASF or P vernatan P of colourchem or any suitable pretanning syntan and 10 per cent water.

Syntan is dissolved in water and added in 3 to 4 installments at an interval of 30 minutes. After the addition of last installment, pelts are further handled with hand in this pit and left immersed overnight.

Malani

9th Day

Malani which is also called colouring pit is the beginning of first step of prior to bag tanning. In one vegetable tanning pit of dimensions 6'x4'x4' which is already in use, the following materials are added.

Mixture of *Babool* Bark and crushed myrobalan nuts in the ratio of 3:1.

25 per cent of the above mixture is added to a vegetable tanning pit containing 300 per cent once used tanliquor. This pit is kept ready one day earlier prior to Malani process.

The strength of vegetable tan liquor in this pit should be about 15 degree barkometer. The leather is placed one by one in this pit and handled with hands continuously for 3 to 4 hours to avoid any sort of oxidation and development of patches so as to obtain uniform colour on the leathers are well immersed in the pit avoiding air pockets between leathers. It may be noted that in the absence of proper handling in the malani tanning pit chances of patches on grain surface of leather are quite possible.

Leather is kept in the malani first tanning pit for 2 to 3 days depending upon the convenience of vegetable tanning.

12ᵗʰ Day

Bag Tanning: Actual bag tanning consists of two parts. In the first part, the leathers undergone preliminary or partial tanning is stitched using locally available date or palm leaves or sisal fibre or moonj. The bag is stitched from butt region of the leather to neck region. The bag of leather stitched on sides is filled with *babool* bark and crushed myrobalan nut mixture in ratio of 3:1, leaving a wide opening near the neck region to pour tan liquor. For three days the tan liquor is poured at regular interval in the leather bag placed over a wooden log. The leather bag is suspended from a wooden log or pole kept perpendicular over tanning pit of size 7'x4'x3' filled with vegetable tan liquor. For three days pouring of vegetable tan liquor is continued avoiding exposure to direct sunlight. By this time the completion of tanning in thickest portion of butt region is checked by knocking with hand giving typical sound.

15ᵗʰ Day

On 15ᵗʰ day the bag is lifted and removed from the wooden pole kept on the base of bag tanning pit on a used gunny bag and neck side opening is stitched with palm or date leaves or moonj and a wide opening is made with the help of sharp steel knife on the base or tip of the butt portion for filling with tan liquor. The leather bag is again hanged on the wooden pole kept over bag tanning pit and tan liquor pouring is continued for 2 days to complete the vegetable tanning in the neck region. After ensuring complete vegetable tanning the bag is removed from the wooden pole cut open to remove tanning materials filled on the flesh side of leather. The used bark is transferred to the bag tanning pit to extract remaining tannin content.

The bag tanned leather is lightly rinsed in the tan liquor available in the bag pit.

17ᵗʰ Day

If necessary bag tanned leather may be bleached to remove any undesirable stains on grain side and further improve the colour into light cream colour with the following chemicals.

Sodium Bisulphite	1 per cent
Water	20 per cent

Applied with a used gunny cloth or salt goat hair brush on grain side and kept in this condition for 30 minutes.

Oxalic Acid	0.5 per cent
Bleaching syntan {Like FCBJ-3 (BASF) if available}	0.5 per cent
Water	20 per cent

This mixture is gently applied on grain surface of tanned leather and kept for 45 minutes to 60 minutes. After this the leather is lightly rinsed in the tan liquor pit.

18ᵗʰ Day

The tanned leather is sometimes kept for 2 days in the bag tanning pit layering with 20 per cent vegetable tanning mixture (Equal parts of babool bark and crushed myrobdan nuts). This tanning mixture is sprinkled on flesh side of leather keeping in pile flesh to flesh and well immersed in the tanning pit. This layering with vegetable tanning mixture further improves vegetable tanning.

20ᵗʰ Day

Leather is taken out from vegetable tanning pit rinsing well to remove adhering bark mixture.

The tanned leather is kept in a pile for 3 to 4 hours either on a flat wooden platform or over a R.C.C. table of size 7'x4'x3', having 5 inch thick top to drain excess of tan liquor.

18ᵗʰ to 20 Day

The tanned leathers are hung in a row close to each other tied by jute ropes inserted in the small holes made on tips of the butt and neck region for conditioning. The room in which vegetable tanned leather are hung may have windows near to roof but there should not be exposure to direct sunlight as sunlight leads to oxidation and darkening of colour of tanned leather. After ensuring the proper conditioning (removal of excess moisture), the leathers are removed from the wooden poles by untying the knots.

Setting

20 day: Leather are hand set on the R.C.C. table described as above with the help of wooden semi curved blunt stone or stainless slicker to remove wrinkles and making the grain surface smooth and leathers flat.

20th to 21st Day

Hand set leathers are again hung in drying room tied with jute ropes and when almost dry are once again hand set to enhance smoothness, flatness and look of the leathers and hung up for complete drying in the drying room. When completely dry the tanned leather is kept in a pile for one to two days for aging.

Yield

Dried and aged vegetable tanned leathers are weighed to find out the yield-percentage. Yield calculated on pelt weight basis may range between 42 to 46 per cent depending on the type of raw material and the method of vegetable tanning.

Vegetable tanned may be sold on piece basis or weight basis depending upon market practices.

In this way good quality vegetable tanning may be completed in 21 days or alternately vegetable tanning in pits may be carried out using 6 tanning pits starting from 10° barkometer to 50° barkometer gradually increasing strength of tan liquor using *babool* bark and crushed myrobalan nuts fortified with vegetable tanning extract procured from tanning chemical suppliers. This is known as pit tanning.

Advantages

- ☆ Employment generating low cost technique at the rural level at the cost of 5-10 lakhs.
- ☆ Environment friendly technology with minimum use of chemicals.

50
List of Some Equipment Manufacturers

1. Swaraj Herbal Plants Pvt. Ltd. Faizabad Road, Barabanki-225001	Aromatic Distillation plants Solvent Extraction plants Aloevera Processing Herbal Extraction plants Fractional Distillation plants
2. Dhopeshwar Engineering Pvt. Ltd. A-16, Co-op. Industrial Estate, Balaji Nagar, Hyderabad-500037	Aromatic Distillation Plants Aloe-vera processing plant Biodiesel plant Floral Extraction plant
3. G.G. Dandekar Machine Works Ltd. Dandekar Wadi, Bhiwandi-42130 Distt. Thane	Rice Mill machinery
4. Kisan Krishi Yantra Udyog, Lalbangla, Kanpur. U.P.	Rice Mill machinery
5. D.P. Pulveriser Industries No.12, Nagindes Master Road, Mumbai-23	Pulverisers

6. Mr. Gunshikhar   Fly ash Brick plant
   89, Bhawathiyar Road, Ganpathi,
   Coimbatore-641006, T.N.

7. Shreya Engineering WorksZone   Dairy Machinery
   B-3, Shed No. C1/309,
   Opp. Old Telephone Exchange
   Anand, Gujarat

8. Varsha Biolers   Boilers
   505, Churchgate Chamber,
   New Marine Lines, Mumbai-400020

9. Mishra Biolers   Boilers
   Industrial Focal Point, Ludhiana

10. Gardeners Corporation   Fruit and Vegetable
    6, Doctor Lane, Near Gole Market   processing plant
    P.B. No. 299, New Delhi-110001

11. Grovers Pvt. Ltd.   Vacuum Shelf Driers
    4-F-4, Shankar Dham, Sundervan
    Complex, Off-Lokhandwala Complex
    Road, Andheri (W), Mumbai-400008

12. B. Sen Barry Co.   Fruit and Vegetable
    65/11, New Rohtak Road,   processing plant
    New Delhi-110007

13. Mangal Engineering Works   Bakery and confectionary
    Factory Area, Patiala-147001, Punjab plant

14. Baker Enterprises   Bakery machines
    23, Bhera Enclave, Near Peera Garhi,
    New Delhi-110087

15. Edwards and Sons   Bakery equipment
    Ambedkar Nagar,
    Deevara Jeevan Halli, Banglore

16. Goma Engg. Pvt. Ltd.   Dairy machinery
    LBS Marg Mezewada, Thane-400601

17. Alpha Level Ltd.   Dairy machinery
    Pimpri, Pune

18. RPM Engineers (India) Ltd.   Small scale milk process-
    14, NP Developed Plots,   ing plant, Ice cream plants
    Thiru Vika Ind. Estate,
    Ekka Thuthangal, Chennai-97

19. Raylon Metal Works Ram Krishna Mandir Road, Kondivili Village, Opp. Mard Bazar, JB Nagar, Andheri (E) Mumbai-400057	Fruit and vegetable processing plant
20. Shanmughan Pillai and Sons 12, Chella DiamamaKoil Street, Dandigul-642019	Continuous roaster
21. Techno Equipments Gali No. 9, Satyam Ind. Estate, Off. Bhaka-Karvi Marg, Behind UVS Limited Govandi, Mumbai-400008	Fruit and vegetable processing plant
22. PWS Engineers Pvt. Ltd. P.B.62, Panchal Estate, Anand-KL Sojitra Road, Anand-388001	Packaging machine
23. S.S Engineering Works A-24, A Bada Mahala, N. Delhi-06	Wafers and snacks machine
24. Central Instt. of Agricultural Engg. (CIAE) Nabi Bagh, Berasia Road, Bhopal	Agricultural machinery

Plastic industries

25. Boolani Engineering Corporation
Prabhadevi Industrial Estate,
402, Veer Saverkar Marg,
Mumbai-400025

26. Kolsitea Machinery Pvt. Ltd.
Veera Desai Road, Andheri (W)
P.B. No. 7368, Mumbai-400058

27. Samarpan Fabrications Pvt. Ltd.
Plot No. A-162/183, Road 16-Z,
Wagle Industrial Estate, Thane-400604

28. Nuchem Plasters Ltd.
H.O. 2016 Mathura Road,
Faridabad-121006

29. SLM Manek Lal Industries Ltd.
Vaswani Mansion Dinshaw Walecha
Road, Mumbai-400020

Index